For thousands of years, the practice of Yoga has demonstrated that we all have the potential to liberate ourselves from much of the ill-health created by the stress of everyday life. Yoga's increasing popularity is based on the growing awareness that positive good health is dependent on an integrated and balanced unit of body, mind and spirit. As a wholistic therapy, yoga treats the whole person, not just the symptoms of ill-health.

M. L. Gharote is a doctor of philosophy with degrees in education. The author of several books on the many aspects of yoga practice and contributor to leading yoga journals and magazines, he is well known to yoga students and teaches in several countries. He is the assistant director of research and the principal of India's Kaivalyadhama College of Yoga and Cultural Synthesis. M. L. Gharote also prepares curricula for university courses and promotes seminars and conferences on research into yoga and preventive medicine.

Maureen Lockhart has been practising yoga since the age of 16, finding it useful in treating chronic bronchitis and, later, a severe spinal injury. A journalist by training, she is also a qualified yoga teacher with extensive experience, and has studied biofeedback, postural integration, humanistic psychology and naturopathy. Maureen Lockhart has edited wholistic therapy publications for several leading London publishers and contributed to journals and magazines. She now lives in India, where she teaches yoga to people of all nationalities and has a private practice in Bombay.

D1380001

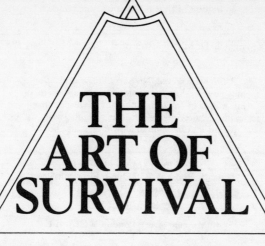

THE ART OF SURVIVAL

A Guide to Yoga Therapy

——Edited by——

Dr M. L. Gharote and Maureen Lockhart

UNWIN PAPERBACKS

London   Sydney

First published in Great Britain by Unwin® Paperbacks,
an imprint of Unwin Hyman Limited, in 1987

UNWIN HYMAN LIMITED
Denmark House, 37–39 Queen Elizabeth Street,
London SE1 2QB

and

40 Museum Street, London WC1A 1LU

Allen & Unwin Australia Pty Ltd
8 Napier Street, North Sydney, NSW 2060, Australia

Allen & Unwin New Zealand Ltd with the Port Nicholson Press,
60 Cambridge Terrace, Wellington, New Zealand

British Library Cataloguing in Publication Data

The Art of Survival: a guide to yoga therapy
1. Yoga–Therapeutic use
I. Gharote, M. L.   II. Lockhart, Maureen
615.8'9   RM727.Y64

ISBN 0-04-149065-7

Set in 10 on 11 point Sabon
by Computape (Pickering) Ltd, Pickering, North Yorkshire
and printed in Great Britain by
Guernsey Press Co. Ltd, Guernsey, Channel Islands.

Contents

Introduction

At 5,000 years old, yoga is the longest surviving fitness fad. It has survived through changes in dynasty, religion and attitude, through wars and famines. It has spread throughout the globe from its birthplace in the Himalayas of India to South America, Russia, Europe, Japan, Australia and the USA. It is practised by millions of people every day in classrooms, homes, hospitals, churches, factories, offices and cultural centres. Yoga has survived this long because it *is* the art of survival. Most of the people who practise it do so because they do not know how they would survive without it.

According to science, the human organism is programmed to survive. The human race is destined to continue no matter what obstacles appear in its path. Even if human society itself becomes an obstacle to survival there will be human beings who will find a way through.

Today, our biggest obstacle to survival is self-induced. We have become the victims of our own unhealthy lifestyle. We are constantly in threatening situations that put our survival mechanisms under pressure so that we live in a perpetual state of stress. Stress-related illnesses are on the increase and conventional medicine is unable to provide satisfactory forms of treatment other than symptom-suppressing drugs and surgery.

In Britain alone around 10,000 people a week are turning away from allopathic medicine and looking for alternatives. As the oldest self-help health system, yoga has increased in popularity while other fitness and health regimes have come and gone.

Why is yoga so popular? The reason is because it is *wholistic*. It helps the whole person become balanced and integrated. Unlike allopathic medicine, it does not treat a symptom on the physical level only, but sees the person as an integrated unit of body, mind and spirit.

In yogic terms, disease is caused by lack of wholeness. Yoga helps us to rediscover how to bring about that state of wholeness within ourselves. Some people are able to do this entirely alone. Some need the help of a teacher to guide them through the techniques. Others are in such an acute state of imbalance that they need the help of a therapist to show them yogic methods to restore health and harmony that are particularly effective for them as individuals.

Yoga is a progressive science, learned in steps and stages. This book is an anthology of essays representing a wide spectrum of experience in the field of yoga therapy, presented in a way that anyone can follow, whether they are a complete beginner, a teacher, a therapist, a health professional or simply an interested reader who would like to know what makes them sick and how they can help themselves become whole and healthy. The chapters progress from the past through the present to the future, giving a comprehensive picture of yoga in its development as a practice; its historical roots, its philosophical foundations, its practical application as a preventive medicine, its efficacy as a therapeutic tool, its role in restoring our natural ability to heal ourselves.

An important issue that this book raises is that yoga therapy as the basis of a self-help medical system would save enormous sums of money currently being spent on research, drugs and equipment, as well as raising the general standard of health worldwide.

Our contributors are drawn from several different countries and from many different fields of activity. They represent a wide range of approaches: from the mechanical to the esoteric; from the known to the unknown areas, yet to be explored by orthodox science but which hold the possibility of our future course of development. Who knows how we may evolve if we can transcend our limitations? If we can go beyond illness, what innate abilities might we all discover that only seem at present to belong to the very few?

For thousands of years, yoga has advocated that we all have the potential to liberate ourselves from our misery. It has evolved the methods and illuminated the way. All we have to do is begin the journey. For some, *The Art of Survival* will be the first step; for others, merely a stepping stone on the way; for all of us, at the very least, it is another opportunity to improve our level of functioning and become truly dynamic, positive human beings.

Dr M. L. Gharote
Maureen Lockhart
February 1987

The Art of Survival

A Guide to Yoga Therapy

PART I

The Premises

1 The Essence of Yoga Therapy

M. L. GHAROTE

Today the ancient discipline of yoga has a worldwide following, and yet there is no field as grossly misunderstood. Although millions of people purport to practise yoga, its essential nature seems to have been grasped by the very few. The problem lies in the fact that yoga is a *wholistic* science whose function is to integrate personality at *all* levels of existence. Hence it is described as unifying, as restoring balance, as inducing homoeostasis or harmony, so the very fact that its various aspects are fragmented and separated into different 'yogas' is thus a contradiction.

Its popularity in recent times is not entirely to blame. Throughout its long history, yoga has passed through numerous changes and developments. India has survived 5,000 years of invasions, wars, cultural practices and differing social views, so we are actually left with very little evidence of how yoga did develop. We are left, in fact, only with fragments, which may in part be the reason why so many minor cults and schools have developed their various different approaches, each emphasising one aspect of the practice rather than the whole. Additionally, each school developed in isolation from each other and from society at large.

However, although the approaches of these various schools differ, their goal remains the same: self-realisation through control of the mind. The literature that remains, namely, the Vedas, Upanishads, Puranas, Smrities, and the texts of Buddhism and Jainism, each describe a different synthesis as the way to achieve that goal. For example, the *Bhagavad Gita* gives the three aspects of jnana, bhakti and karma, based on the three faculties of man: intellect, emotion and physique. The *Yoga Sutras of Patanjali* describe a systematic, eightfold development of practices graded progressively. Whether the particular school employs the techniques of hatha yoga, karma yoga, raja yoga or laya yoga, the goal is always the same for, traditionally, spiritual liberation was seen as the only desirable goal worth attaining. This emphasis on a particular set of techniques to achieve that goal has led to confusion about the purpose of yoga. With the discovery that yoga has a therapeutic effect, especially on stress-related illnesses, the spiritual goal has been subsumed by the need for physical fitness. In fact, the therapeutic aspect of yoga does not feature in any of the traditional systems of self-help, except in the *Yoga Sutras of Patanjali* where we come across the word vyadhi meaning 'disease' in the list of disturbing factors of mind that are obstacles to liberation.

In yogic terms, a 'dis-ease' or imbalance which cannot be corrected by right living, right eating, right thinking and meditation on the spiritual goal requires the intervention of a specialist, and for this purpose the science of ayurveda evolved. However, ayurveda also recognises the goal of liberation and accepts the contribution of yoga towards its attainment. Furthermore, the therapist is regarded as one who shows the aspirant the path of liberation through a daily routine of self-discipline based on yoga practices, through the principle of moderation in food, exercise and behaviour and through the use of yoga practices to provide relief from all kinds of physical and mental pain. Both systems consider the concept of prana to be of vital importance in their respective eightfold paths.

Although yoga therapy was not a developed branch of yogic discipline as such, we do get a glimpse of the therapeutic effects of the practices in some of the hatha yoga literature such as the *Hatha Yoga Pradipika*. However, advice is given here within the context of practice; that is, how to deal with the complaints that arise from faulty practice, such as chest pain, pain in the sides, backache, headache, asthma, hiccup, skin disease, partial blindness, tremors, blood infection, and so on. The terms used are ayurvedic rather than yogic; for example, the text suggests that 'Whenever any region is afflicted by disease one should contemplate upon the vayu situated in that region' and take the help of ayurvedic medicine if necessary.

Looking at yoga from the context of our own lives, if we consider the word 'therapy' in its broadest sense to mean 'correction of the abnormality and leading to the ideal state of normality expected in yoga', then the whole field of yoga is nothing but therapy. Any state of deviation from the ideal condition is a state of disease – a lack of harmony – and to re-establish this harmony through yogic techniques is yoga therapy. Improved communications in the twentieth century have enabled the living science of yoga to continue its growth and development. Part of that growth has been the integration and exchange of various features from the many different cultures into which it has spread throughout the world. When the Occident came in contact with the Orient through commerce, it exported more than spices and silks, for Western people were exposed to indigenous arts and sciences. This was especially the case during British rule in India, for British administrators took a keen interest in all aspects of Indian life.

The study of Indian philosophy began in some of the European universities as early as the eighteenth century. Among the German philosophers were Schopenhauer, Schlegel and Paul Deussen. Many scholars studied Sanskrit and translated Indian classics, among the best known being Max Müller and Sir John Woodroffe, who translated the Vedas, the Upanishads and some of the tantras. Interest in yoga was stimulated in those early days by visits to America and Europe of Raja Ram Mohan Roy, Keshab Chandra Sen, Govindananda Bharati and Swami Vivekananda, and in more recent times of Sri Aurobindo, Ramana Maharshi and Paramahamsa Yogananda. Undoubtedly, the patronage of the Beatles helped to spread the Maharshi Mahesh Yogi's TM meditation, and the concomitant proliferation of health magazines, in which pop singers, movie stars and TV personalities feature in yoga poses, have had a more widespread and far-reaching effect on the present young generation than those pioneer writers Paul Brunton, Alexandra David-Neel and Theos Bernard did on their parents and grandparents. However, there are many worthy organisations and institutes, both inside and outside India, like the Theosophical Society and the Sivananda Vedanta Yoga Centre, that have encouraged the exploration of yoga beyond the superficial aspects of fitness and beauty. Yoga today is a serious subject for research by scientists and medical researchers.

Much of the pioneering work in this field was done by Swami

Kuvalayananda at the institute founded by him at Kaivalyad-hama, Lonavla, some 80 miles from Bombay. Since he began this work in 1924, his experiments to study the benefits of yoga practices have been reproduced numerous times in laboratories in several different countries. Perhaps the most valuable contribution made by these experiments was the birth of biofeedback, a true integration of the East and the West, of the old and the new, through the marriage of ancient wisdom and modern technology.

The most immediate and relevant contribution of yoga is as an antidote to stress. There are a multitude of yoga techniques which work effectively in removing psychophysical tensions, especially that particular feature of hatha yoga practice – stretch. But although there are many individual techniques and approaches to deal with stress, it is its *wholistic* approach to the art of survival through self-help that makes yoga therapy so effective. When the physical, mental and spiritual aspects are integrated, the preventive, promotive and curative qualities are allowed to develop, and each of us thus has the opportunity to evolve and heal ourselves.

Yoga therapy has a long way to go. To understand its essence the mechanisms involved in yoga techniques need to be fully explored through scientific and clinical research. We need to know more about how the organs of the body, the muscles, joints and ligaments are affected by the act of stretching in asanas; we need to understand how pranayama, bandhas and mudras affect the balance between the sympathetic and parasympathetic nervous systems; we need to learn how to establish voluntary control over the somatic system, as well as over the psychic areas of the brain, through meditative practices; we need to study methods of producing natural mood-altering chemicals through change of mind. We need to know more about pain, and about illness as a whole: What does it do to us as people? Does it have a function in shaping our characters – or our destiny? We need to know more about energy, about the source of life itself.

Perhaps it will be many years before we fully understand what man is all about. Perhaps it will take many more experiments, both in the laboratory and within ourselves, before we can unveil the mystery of our vital essence – our Selves. But in the mean time, all of us who work in the field as teachers, therapists and researchers continue to contribute to the science of life in the spirit of yoga.

2 The Experiences of Yoga Therapy

MAUREEN LOCKHART and
PETER V. SANDHU

In just two decades yoga has become a household word in Britain. Twenty years ago, if you talked about the experience of yoga with someone who did not practise it (and very few did) you were thought 'weird'. Ten years ago you were a 'health freak'. Today it has become an acceptable part of the national way of life for thousands of people, and an important part of the wholistic health movement. One of the most important factors about the current interest in wholistic health systems is that we have finally got to talk about ourselves as we really are; it has opened up a whole new area of communication, especially with those who specialise in taking care of our health problems. Yoga, as the oldest system of wholistic therapy, has made a valuable contribution to this communication through its systematic and scientific methods based on 'subjective' experience.

The wholistic age in which we live has provided us with the opportunity to talk more freely about what we observe, what we feel and what we *know* to be going on inside us, in a language that is natural to us. Lay people do not talk in medical terminology, a fact that has been frequently overlooked by those trained to think in medical terms. However there are still great gaps in communication that need to be healed. The reductionist medical model has not only narrowed the physician's view of the patient but has also severely limited the patient's ability to talk about his condition, with disastrous results. Thus people often do not seek help until they are in the last stages of disease, when the illness has visibly manifested itself and has wreaked havoc on the organs and structures of the physical body. Because the patient is afraid to bother an overworked doctor in an overcrowded clinic with symptoms that are subtle, and hence 'trivial', he tends to wait until his symptoms become overt. This encourages a curative rather than a preventive approach to medicine and invariably results in long-term and expensive treatment in which the patient is not a partner in the process but merely a passive recipient.

This change in the relationship between doctor and patient came about with the advent of the 'pharmaceutical age'. Until the 1950s, doctors relied as much on their 'bedside manner' and their ability to counsel patients as they did on the small range of drugs that were available to them. People also treated themselves for minor ailments with folk remedies that had been tried and tested through numerous generations, and only sought help when symptoms did not respond to those treatments. Specialisation, together with antibiotics and other antibacterial drugs, changed all this. A new age of 'scientific' medical practice was ushered in, in which the patient became dependent on multipurpose medicines. The rate at which literature on these new permutations of 'superdrugs' was produced made it impossible for most doctors to keep up, and people who did not respond adequately to drug therapy were often referred to psychiatrists. In effect, the doctor had handed over his counselling role to a specialist, and although a certain number of people have been effectively helped by psychiatry, the emphasis on mental disease negated the body and led to another kind of polarisation. In addition, dependence on treatment by drugs, particularly in hospitals, led the profession away from Freud's – and later Jung's and Reich's – original broad-based (we would now say wholistic) therapy, until the advent of humanistic psychology which emphasised a total body–mind approach.

The study of human potential and growth in the normal person, rather than the sick person, led to the exploration of human consciousness and methods of helping people to help themselves without drugs. In fact, humanistic psychologists were among the first to make people aware of the dangers of drug-based therapy and shock treatment, and advocated instead natural methods like yoga and meditation, together with counselling, psychodrama, sensory awareness training and a variety of other techniques in personal exploration. For the first time, the patient was regarded as having an active role in the healing process and therefore his *experience* was important. In contrast, the scientific-medical model reduced the person to an object; his symptoms were regarded merely as 'soft data' that had to be backed up by the hard facts – the laboratory tests, X-rays, etc., that emerged from the technological hardware that the medical profession now find indispensable.

This has led to a split, not only between doctor and patient, but between the patient and himself. Even if he feels well, he cannot be sure that he is well until the tests come back from the lab. He has no faith in his own knowledge of himself. He does not trust his own perceptions and gradually he fails to perceive and depends totally on the doctor. This is an unhealthy situation, for the more people like this the doctor has to treat, the more overworked he becomes and the less time he has to give to each patient; hence he has to rely on outside sources in order to confirm his hasty diagnoses.

There is only one solution to this problem, and people *are* being attracted to it. The experience of yoga, sought out by both dissatisfied patients and doctors, has brought in a new feature – self-help based on the efficacy of experience. And this may be the single most important factor in the health business, for experience has been vastly underrated. As Thelma Moss admits in her book, *The Probability of the Impossible*, when talking about experience of the paranormal: 'Scientists have never been satisfied with anecdotal material however convincing the story or unimpeachable the source.'[1] Unfortunately, this has been true about normal experiences, too. People who have turned to yoga, often out of desperation when they cannot find help for a chronic complaint, tell numerous distress stories about the disregard some specialists have for the person's right to own their own bodies and to have a say in their treatment. Patients have reported stories about being asked, when they dared to venture an observation about their condition: 'What do you know about your body; are you medically trained?'

For example, one young woman we know had had a fall, after which she suffered acute pain and dizzy spells and felt that her body was becoming increasingly asymmetrical. One of the several doctors who had looked at her, and found nothing 'wrong' with her, informed her that it was 'all in the mind' and suggested she have psychiatric treatment. She was acutely distressed by this 'diagnosis' since, as a student of yoga, she knew that there was a serious discrepancy in the *feel* of one side of her body, which behaved differently from the other side when she was practising asanas. Even after the condition deteriorated, to the extent that one side was periodically paralysed, she was unable to obtain help. In desperation she contacted a well-known healer who, convinced that she would respond to 'structural treatment', referred her to her own chiropractor.

In retrospect, the thing that impressed this woman most about the chiropractor was his way of communicating. He listened very attentively to what she said and encouraged her to tell him what she felt was 'wrong with her'. Then he explained what her feelings meant in structural terms. Her pelvis had tilted 12 degrees to one side, which was why she had the feeling of being asymmetrical – she *was* asymmetrical and her body was telling her so quite clearly. Later on, during the course of treatment, she began to have dreams of being pregnant. When tests proved that she evidently wasn't, she reported this experience to her therapist, having been encouraged to speak freely about anything she observed, no matter how trivial it seemed. This, he explained, was a very good example of the mind picking up information from the senses and translating it into primary process language. To illustrate what he meant by this, he showed her on a skeleton the shape that her injury had forced her spine to take – which was exactly that same shape that develops during pregnancy. Hence her dreams of being pregnant. The mind had translated her sensory impressions and fed them back to her, quite literally. Until one learns to read and understand this language, it can be very confusing for the person getting the messages.

Body–mind therapists stress the importance of understanding this form of communication. As Will Schutz and Evelyn Turner point out, 'degrees of self-awareness are manifested both psychologically *and* anatomically'.[2] A loss of integrity or integration in the human system, whether it originates from within as a particular way of thinking or from without as a physical injury or postural habit, affects the *whole* person. 'When a body is traumatised, not just "one" thing happens, but the body as a whole is involved, everything, everywhere.'[3]

Furthermore, 'traumas of sufficient force'[4] can permanently change the shape of the body. The trauma can constitute anything from an emotional shock to persistent pressure through hard physical work, or even uncomfortable clothing. The constant use of the body in a particular way, for example the ballet dancer, may produce a pattern of chronic muscular tension that is as painful as the pattern produced by acute anxiety.

Yoga has always recognised the fact that structure and function are intimately related; hence the wholistic approach to the person through a repertoire of techniques that retrain body, mind and feelings as one entity. Students of yoga become more aware of sensory phenomena, which they learn to interpret correctly and make subtle adjustments in mental, emotional or postural attitudes. In her essay on hatha yoga in *Rediscovery of the Body*, Rama Jyoti Vernon calls this ability 'clairscience', which, she says, is related to the heart chakra (energy centre) traditionally related to the sense of touch or feeling.

> As the physical nerves and the plexes have a close relationship with the chakras, it is believed that vibrations produced in the physical centres have similar effects upon the subtle ones. Thus, the intensification of vibratory currents produced in the physical centres by hatha yoga is said to have a correlative effect upon the chakras. As each chakra is a storage place for energy forces, it represents a state of consciousness. As this consciousness manifests as energy, and energy as consciousness, hatha yoga directs the student towards increasing one (energy) to heighten the intensity and sensitivity of the other (consciousness).[5]

The development of 'clairscience' is what allows the true therapist, in whatever field, to communicate from the heart. When the heart is open, a more subtle level of communication between therapist and patient is possible. Unfortunately, experiences of this nature have not attracted much interest among scientific experimenters. The much rarer and more dramatic gifts of clairvoyance and clairaudience have claimed scientific attention today, in much the same way as the study of abnormal or aberrant states of mind did in a previous era, while clairscience, which would seem to have much more potential for human development among 'ordinary' people – and which may even be a vital factor in understanding clairvoyance, clairaudience and other so-called 'supernatural' states – has hardly been explored.

Undoubtedly, this has something to do with the scientific process itself and, more specifically, the mechanistic (non-feeling?) approach. Meditators who have taken part in scientific experiments have often been bemused by the scientists' view of the meditative process and the conclusions drawn from the study of selective phenomena. For example, a psychiatrist who was studying brainwave patterns in meditation for the purpose of comparing them with the patterns of mentally disturbed people, concluded that the feeling of euphoria reported by meditators (and usually referred to as a 'heightened state of consciousness') resulted from the 'idea of well being' that came from 'anticipation'. One of the experienced meditators on the project pointed out to him that this may make 'logical' sense to people trying to understand something they have no personal experience of from the so-called 'objective' point of view (provided they could even explain to themselves how one can anticipate the repetition of an experience unless one has already had it), but from the 'subjective' viewpoint it was abject nonsense.

The meditator then attempted to explain to the scientist that the phenomena he was studying, i.e. alterations in brainwaves, did not exist in a vacuum, for while he may have elected to study that particular phenomenon, he had isolated it from a whole pattern of other phenomena that the meditator was experiencing. The euphoria was not an 'idea' in the meditator's head, but a *feeling* throughout the *whole* of himself. Furthermore, it was part of a *process* that was brought about by the control or manipulation of certain other felt sensations, that had outward manifestations as well as inward ones. The scientist had evidently failed to notice that as the meditator went more deeply into meditation – as his brainwaves changed from alpha to theta – his body was also going through a simultaneous structural change; his spine was spontaneously straightening itself, a phenomenon that is of central importance in yoga practice for many reasons.

Somewhat crestfallen, the scientist asked if he should redesign his experiment. 'No', replied the meditator, 'redesign your life. Learn to meditate yourself.' And to his great credit, the scientist did learn to meditate.

When they next met, nearly two years later, and the meditator asked the psychiatrist what difference meditation had made to his work, he replied that it had made 'a world of difference'. Having the first-hand experience of meditation made him realise that he was asking all the wrong questions, and the conclusions he had drawn from his experiments had no value because they were based on a false premise.

There are numerous phenomena in the yogic experience that have a practical application in the field of therapy. The whole of yoga, in fact, represents thousands of years (and perhaps billions of experimental hours) of experience. This knowledge has been organised into a system of practical methods that anyone can follow under the guidance of someone who has 'been there'. Does this mean that the therapist or doctor has to have experience of every disease that their patients suffer from? Definitely not. There are numerous experiences that are universally common, that are the basis of the 'human experience'. However, what the doctor or therapist does require is 'clairscience', to remind him/her that he/she is not separate from the patient but that we are all one, we are *all* part of the human condition. We need to be able to feel with sensitivity, to hear and see clearly what the patient needs to help him become whole. And in order to communicate that effectively we need to practise awakening all our faculties. We need to understand the language of experience, our common bond in the search for a wholistic and healthy way of life. It has been language, after all, that has enabled us to take a gigantic step forward in our evolution; that has enabled us to apply what we have learned from personal experience in making our knowledge of our world work for us; that has enabled us to say 'Physician, heal thyself' and understand what that *really* means.

REFERENCES

1 Thelma Moss, *The Probability of the Impossible*, Paladin, 1979.
2 Will Schutz and Evelyn Turner, 'Bodymind', in *Rediscovery of the Body*, ed. Charles A. Garfield, Laurel, New York, 1977.
3 Robert Prichard, 'Structural integration,' in *Rediscovery of the Body*, ed. Charles A. Garfield, Laurel, New York, 1977.
4 Robert Prichard, 'Structural integration', in *Rediscovery of the Body*, ed. Charles A. Garfield, Laurel, New York, 1977.
5 Rama Jyoti Vernon, 'Hatha yoga', in *Rediscovery of the Body*, ed. Charles A. Garfield, Laurel, New York, 1977.

3 The Principles of Yoga Therapy

M. V. BHOLE

Although yoga was not originally developed as a system of therapy in the strictest sense of the term, it is becoming accepted as such. The high cost of drugs, their toxic side-effects, problems in the doctor–patient relationship, have all stimulated people into looking for a natural system of therapeutics that not only avoids the use of harmful drugs, but really does tackle what one might call the 'existential' problems of being human.

In fact, it is only recently that these existential problems have been acknowledged, for science did not have an appropriate investigative framework for looking at so-called 'subjective' states. However, new developments in physics and the behavioural sciences have made it possible to understand the ancient science of yoga, its approaches, practices and principles, and in so doing have clearly added a new dimension to medical science.

In yoga it is believed that pure consciousness is non-material and therefore different from the body (structure), mind (feelings and emotions) and intellect (thinking and decision making). This consciousness is said to work through an energy or force called prana. Therefore, any imbalance or disharmonious functioning of the body, mind and intellect is taken to represent an imbalance of the working of the prana. Judicious use of various yoga techniques are found to correct imbalances in the working of prana, which in turn will be reflected in the body, mind and intellect. This can be taken as the basis of yoga therapy and with this understanding I shall now proceed to discuss the principles involved in greater detail.

From the moment of conception there are constant interactions between oneself and the environment. Some of these interactions help in the positive growth of an individual, promoting physical, functional, mental, moral and intellectual development, while others have a negative influence. During such times, various adaptive mechanisms are developed, in the interest of survival. Survival becomes smooth if these mechanisms are adequate and efficient: if not, 'mal-adaptation syndromes' develop. These are manifested at various levels of human existence, at different phases and times during the lifespan. They come in the form of a general feeling of 'dis-ease', 'dis-turbance', 'dis-harmony', 'dis-integration', 'dis-equilibrium' and/or 'im-balance'. They have different symptoms and signs which are initially vague and then more localised to different parts of the body, mind and intellect. Symptoms are thus the abnormal subjective sensations and signs that are the objective data which can be verified by a therapist. Hereditary factors or earlier traumas and 'insults' to the organism make some people more sensitive or susceptible than others to certain circumstances.

As far as the applications of yoga are concerned, they can be employed to advantage in chronic and sub-acute disease where the psycho-neuro-endocrino-muscular system is not functioning properly or adequately. However, acute infections, accidents, malignancies, hereditary disease related to aberrant immune body-responses and mental derailments may not respond to yoga therapy and therefore one should not take it as a panacea for all sorts of human sufferings.

THE GENERAL AND INDIVIDUAL APPROACHES

Two approaches emerge in the practice of yoga therapy: generalised; and individualised. Let us look a little more closely at these approaches and see what techniques are most appropriate.

In the general approach the most important aims are to help the patient:

1 relax properly so that tensions can be released;
2 be aware of certain mechanisms at work in the body, like sense of balance, body awareness (proprioceptive

awareness), awareness of internal organs (visceral awareness), so that these imbalances can be altered through suitable techniques;

3   experience certain functional planes and axes in the body;
4   become aware of breathing and remain in that awareness for some time in a relaxed way;
5   observe various thought processes without suppressing them or getting involved in them;
6   have the inner experience of self-dependence, in order to reduce dependence on external objects by a gradual process;
7   learn to practise each technique with full awareness of the inner responses (for this reason practising asanas and pranyama with eyes closed is advisable).

The individual approach, however, helps the patient to become aware of:

1   tensions and conflicts present in the body, and how to release them in an efficient manner;
2   the breathing pattern, to correct it if it is wrong and how to employ it in the practice of asanas;
3   posture, equilibrium and the nature of muscle tone, and how to restore these through asanas;
4   the sequence of events and the importance of each event in natural breathing and how to work with it through various breathing techniques;
5   feelings and emotions, tensions and conflicts, and how to deal with them through suitable practices;
6   personal value system and attitudes to life as a whole, and how they need to be altered in the interests of the self and society at large;
7   the role of various environmental factors like living conditions, working conditions, food, etc., in the state of imbalance and what remedial measures can be taken;
8   the possibility of experiencing a state of undisturbed, well-balanced self-consciousness and learning to maintain that state through suitable techniques.

The next point that we have to consider is just what exactly are those techniques that have been referred to and what makes them important therapeutic tools? There are in fact many suitable techniques, but here I propose, confined as I am by limitations of space, to look only at three important groups: asanas, pranayama and kriyas.

ASANAS

The term asana literally means a 'seat' (chair) and as such is used by a person to sit in or sit upon (like a carpet). In physiological language, it could be translated as a 'posture' or, better, a 'postural pattern'. Unfortunately, though, many yoga teachers tend to look upon asanas as exercises, and so the real therapeutic value of postures is lost. Even medical and paramedical people fail to understand this important concept, so the whole evaluation of yoga has been wrongly based on the assumption that it is an exercise system. In practice, the seat or asana and the person occupying the seat or asana are two different entities.

Let us now try to understand this more clearly by looking at the difference between muscular exercise and yoga asanas. Medical physiologists differentiate between the physiology of muscular exercises on the one hand, and tone, posture and equilibrium on the other. The motor cortex, that is will and volition, is more involved in muscular exercise, which is closely connected to the voluntary and chosen behaviour of the person. The cerebellum, however, is more involved in the maintenance of tone, posture and equilibrium, which are closely related to an individual's involuntary and emotional behaviour.

For the practice of *pranayama* and meditation yoga scriptures recommend a balanced, straight, but not erect, and relaxed condition of the body, particularly the trunk, thorax, neck and head. However, few people seem to be able to attain this ideal posture because of imbalances at the physical and mental levels. These are manifested on the physical level as hyper-flexion or hyper-extension, i.e. hypo- or hyper-tonicity, and could be corrected by *asanas* such as those shown in Figure 3.1.

Some of the salient features of asanas which differentiate them from muscular exercises are as follows:

1   Certain groups of muscles are stretched while other muscles are relatively relaxed.

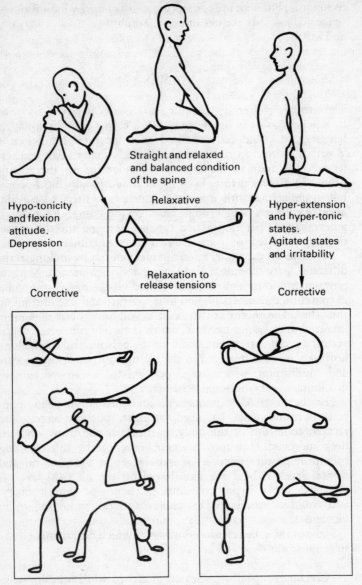

Straight and relaxed
and balanced condition
of the spine

Hypo-tonicity
and flexion
attitude.
Depression

Relaxative

Relaxation to
release tensions

Hyper-extension
and hyper-tonic
states.
Agitated states
and irritability

Corrective

Corrective

Figure 3.1   Asanas to correct hypo- and hyper-tonicity.

2 This enables various joints to receive exercise through the entire range of possible movements.
3 The focus of the postures is on the spine and so the long and short muscles of the vertebral column through the spinal nerves receive maximum attention.
4 The internal viscera are stimulated and maximum effect on the visceral system can be further attained by breathing practices.

PRANAYAMA

The subject of pranayama[1] is greatly misunderstood. Most books merely equate it with a breathing exercise for better gaseous exchange, the improvement of lung ventilation and blood oxygenation, though laboratory investigation does not seem to support this.

Prana is regarded as the life energy or life force of the individual, at work on various levels of existence. Breathing is one of the physical manifestations of prana which can be voluntarily modified, and so it is primarily on this level of breathing, in spite of the use of cleansing exercises which also affect the prana, that the experience of prana is best understood.

And yet very few people either really experience their breathing or are observant about changes in the breath, however much they think they know about the functioning of the respiratory system. We are here more concerned with the awareness of breathing available to us, firstly, through physical movements at the periphery (surface) of the body and, secondly, through the touch of air felt inside the passages. Placing hands over different areas of the chest and abdomen as shown in Figure 3.2 while taking two or three normal and deep breaths can give the first type of awareness. It will tell whether or not:

1 only the front part of the chest moves during breathing;
2 only the front part of the abdomen moves during breathing;
3 the sides of the chest and/or the abdomen move slightly, or not at all;
4 the movements in the right and left sides are equal or unequal;
5 the back is completely inactive or only partially active;

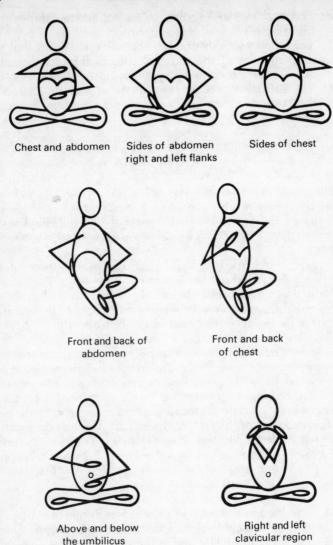

Chest and abdomen

Sides of abdomen
right and left flanks

Sides of chest

Front and back of
abdomen

Front and back
of chest

Above and below
the umbilicus

Right and left
clavicular region

Figure 3.2 Experiencing breathing movements in various areas of the chest and abdomen, with the help of the hands.

6   the abdomen goes in while the chest comes out, and vice
    versa;
7   the sides go in while the abdomen/chest come out, and vice
    versa;
8   the back goes out when the abdomen goes in, and vice versa;
9   the part below the umbilicus, extending to the perineum,
    moves in and out;
10 the clavicular area is taking part in breathing activity.

The oxygen consumption, CO_2 output, minute ventilation, etc.,
remain almost equal in all these conditions, so we can therefore
describe this as the breathing pattern of an individual. The
question then arises: What is the ideal and correct breathing
pattern? In short, the simultaneous expansion of the chest and
abdomen in *all* directions during inspiration and the simultaneous shrinking or contraction of the chest and abdomen
during expiration is the ideal breathing pattern, which few people
seem to have. Furthermore, a person who is breathing correctly
should be able to feel the breathing movements *all over the body*,
from the tips of the toes to the top of the head, strange as that may
sound. Therefore, it is a fundamental part of yoga therapy that
the breathing pattern should be corrected before moving on to
any other *pranayama* practices, and this in itself will provide
immense benefit by releasing tensions and correcting imbalances.

Trying to be aware of the touch of air in the nostrils is the
second approach to understanding one's breathing can give the
following type of information:

1   unable to feel anything within the nostrils;
2   able to feel the touch of air over the lip and brim of the
    nostrils;
3   able to feel during inspiration but not in expiration, or vice
    versa;
4   able to feel the touch of air in only one of the nostrils;
5   experience of temperature changes or movements of the
    nostrils instead of the touch of air;
6   able to feel the touch of air beyond the nostrils.

With the help of various cleansing processes and pranayamic
techniques[2] one can become sensitive to the touch of air within
the respiratory passages and can learn to organise these

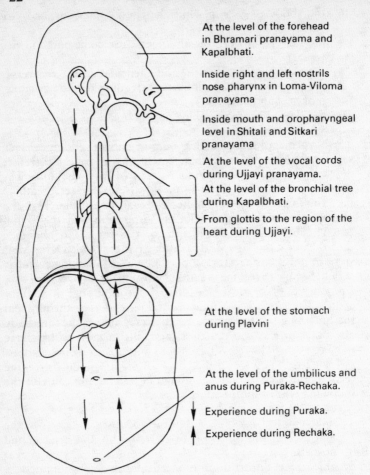

At the level of the forehead
in Bhramari pranayama and
Kapalbhati.

Inside right and left nostrils
nose pharynx in Loma-Viloma
pranayama

Inside mouth and oropharyngeal
level in Shitali and Sitkari
pranayama

At the level of the vocal cords
during Ujjayi pranayama.

At the level of the bronchial tree
during Kapalbhati.

From glottis to the region of the
heart during Ujjayi.

At the level of the stomach
during Plavini

At the level of the umbilicus and
anus during Puraka-Rechaka.

Experience during Puraka.

Experience during Rechaka.

Figure 3.3 Experiencing the touch of air within the respiratory passages, through various yoga techniques.

sensations and relate them to various other happenings initiated by the breathing process (see Figure 3.3). Thus one learns to breathe completely, ensuring that the whole body cavity is filled (puraka) during inspiration (svas) and emptied (rechaka) during expiration (prasvas) and is either filled or emptied in the suspended state (kumbhaka or shunyaka) in the pause between inspiration and expiration, either voluntarily or autonomically in different stages.

Unlike the *Hatha Yoga Pradipika* (a yogic text written in the fifteenth century), Patanjali's *Yoga Sutras* (written around AD 200 and now regarded as 'classical yoga') does not refer to the terms puraka-kumbhaka-rechaka, but only mentions tackling svas-prasvas in a particular way. However as respiratory movements are dependent on nervous impulses reaching the periphery of the body from the brain, any successful manipulation of the breathing pattern is going to result in more efficient nervous transmission. Thus many positive functional changes can be implemented through certain carefully selected pranayama practices.

KRIYAS

There are only six purificatory practices mentioned in *Hatha Yoga Pradipika*, while another classic work, *Gheranda Samhita*, mentions many more. These practices are supposed to be complementary to pranayama. Even a superficial study of them reveals that they are related to those internal organs of the body that are mostly governed by the autonomic system and which are thus highly susceptible to emotional disturbances in their functioning. Therefore, a positively motivated method of working with kriyas helps to modify and monitor the functions of the neuro-vegetative system and hence the emotions and behaviour.

1 Neti kriya is especially helpful in chronic conditions of the nasal passages and air sinuses and helps a person feel and evaluate the touch of air in the nasal passages during pranayamic breathing.
2 Kapalabhati improves breathing through the working of the abdominal and thoracic muscles, increasing awareness of the touch of air in the bronchial tree and the forehead. The strong expiration expels bronchial secretions and so clears the respiratory passages.
3 Dhautis help to increase awareness of the oesophageal passage and can be used to turn negative responses into positive ones, so affecting behaviour. Generally, vomiting is considered a repulsive or disgusting act and the idea of putting

a rubber tube down the throat or swallowing a piece of cloth is very off-putting. But one becomes so used to it and enjoys the after-effects of cleanliness so much that it becomes a pleasant experience.

4   Gajkarni and vamana dhauti are methods of swallowing water to wash out the stomach. Here again, the unpleasant act of throwing up is converted into a pleasant experience of feeling cleansed, and increases well-being.

5   Nauli gives mastery over the abdominal muscles, massaging the intestinal tract and so improving digestion.

6   Basti, with the help of nauli, washes out the colon keeping the system free of toxins. The experience of water rising in the colon sensitises the student and prepares him for certain mudras and bandhas.

In addition, tratak (steady gazing) to the point where the eyes water is helpful for the health of the eyes and also purges the mind of projected images in the second stage when the eyes are closed. This has a positive psychological effect.

Shankha-prakshalana has been found to be very useful in the management of various functional disorders, partly through the stimulation of the autonomic nervous system and partly through mechanical factors at work during this practice, i.e. peristaltic action can be set in motion and the contents of the whole intestinal tract evacuated and cleansed within 1½ to 2 hours.

As will now be clear from the above discussion, the most important aspect of these particular yoga therapy practices is to increase sensitivity to internal phenomena and thus provide a more efficient sensory feedback system. This results in speedier reactions in the interests of survival. For example, if a person eats a food that is not fresh or to which they may be allergic, they will have a chain of reactions which may result in ill health through the interrupted functioning of the body. However, a student of yoga who is adept in at least some of the above practices will either automatically throw out that alien substance or will receive clear signals that they should initiate rejection of the substance, thus preventing the unnecessary loss of vitality that would otherwise result.

Gradually, through consistent practice and experience, the yoga student learns to utilise a wide range of techniques to increase awareness on many levels. In this way, he takes control of his own existence and becomes responsible for his own health and well-being.

REFERENCES

1   Swami Kuvalyananda, *Pranayama*, Popular Prakashan, Bombay, 1972.
2   Ibid.
3   Swami Prabhavananda and Christopher Isherwood (trans.), *How to Know God: The Yoga Aphorisms of Patanjali*, Mentor.

4 *The Empirical and Experimental Foundation of Yoga Therapy*

CTIBOR DOSTALEK

The scientific and technical achievements of modern society bring with them new possibilities for personal growth, but they also give rise to a complicated stressful milieu for man – increased demand on adaptive processes in order to manage the wide spectrum of stimuli which, together with environmental pollution, form the basic causes of the dangerous increase in the number of so-called civilisation diseases.

These phenomena have led to increased use of medicines, especially synthetic tranquillisers and hypnotics, most of which have very marked side-effects. This fact is partially responsible for motivating the medical sciences to search for therapeutic approaches which are not only anoxious but which also reinforce the body's natural capacity to resist disease. Given these circumstances, the widespread interest in yoga is natural, especially since the emphasis of yoga is on preventive rather than curative measures. This emphasis is in accord with the new medical model, which is both wholistic and therapeutically more humane. It is therefore becoming more common to find the medical aspects of yoga on the agenda of congress topics in both the East and West.

Yoga in its broadest sense covers the whole spectrum of human activity, but from the point of view of medical goals it is perhaps more useful to concentrate on hatha yoga, since this can be described as the somatic aspect of the integrated yoga process. Hatha yoga may thus be defined as an empirical system of exercises leading to the stabilisation or homeostasis of the human regulatory processes. Its practices are both preventive, increasing the capacity for resistance, and curative and/or rehabilitative. Hatha yoga ranges from very easy exercises, which even the seriously ill can attempt, to quite advanced practices requiring lengthy preparation. Furthermore, hatha yoga includes both relaxation (inhibition) and activation (excitation). The end result of both is more efficient homoeostasis of the regulatory processes.

RELAXATION

Savasana (corpse pose) is probably the best known relaxation exercise. Through relaxation of muscle tensions, an anxiety-free state is reached. Savasana has therefore proved to be a suitable tool in managing hypertension[1], even in cases that are unresponsive to drugs. In his book, *Disorders of Stress and Their Management by Yoga*, K. N. Udupa[2] reports that in one experiment with hypertensive patients conducted by Agrawal, Vaish and himself, 65 per cent of the patients not on drugs showed a significant improvement after practising 30 minutes of savasana daily for three months, while 80 per cent of the patients in another group responded positively to savasana in combination with drugs. Udupa also shows that savasana is effective in functional cardiac disorders. Furthermore, in experiments conducted with Singh and Dubey over six months he discovered that it has similar effects to diazepam, which is also an effective muscle relaxant leading to freedom from anxiety. Thus savasana can be a valid replacement for synthetic drugs which have harmful side-effects.

Still on the subject of heart problems, it is worth mentioning briefly that sirsasana (head stand) has an impressive effect when performed at an advanced level (though it evidently also has beneficial effects when performed by less advanced students). Well-known yogi and author, Theos Bernard, practised sirsasana 3 hours each day (1 hour three times daily) which resulted in the lowering of his heart rate to 12 beats per minute and his breathing rate by three to four cycles per minute when at rest.

Nauli was found to be one of the most interesting techniques in this respect. Intense practice of nauli can slow the heart (brady-cardia) by up to 33 cycles per minute and ventricular changes can be demonstrated at the end of the practice. This is in accord with Karambelkar and Bhole's[3] findings of the effect of three bandhas on the heart rate.

Pranayama is viewed as the most important of the hatha yoga practices, and physiological measurements back up this tradi-tional belief. The most striking cardiovascular changes were found mainly in the retention phase of the breathing cycle, called kumbhaka. These changes were more prominent during retention after expiration. Kumbhaka performed immediately after kapa-labhati effected a prolongation of the PQ interval of the ECG by up to 300 milliseconds.[4] Advanced practice of sitali produced significant changes in the blood pressure and kumbhaka with jalandhara bandha and mula bandha lowered blood pressure from the resting value of 105/50 mm Hg to 70/35 mm Hg.[5] Nadi sodhana pranayama created an even more significant drop in blood pressure.[6]

Experiments of this sort have proven that the popular expla-nation of the effectiveness of pranayama due to so-called 'better ventilation and oxygenation of tissues' is quite wrong. Hyperven-tilation at rest in fact causes loss of CO_2 (hypocapnoea), the result of which is cerebral vasoconstriction, producing ischaemic anoxia.[7] Besides, during hypocapnoea haemoglobin releases less O_2. In contrast, in pranayama voluntary apnoea and extension of the expiration is emphasised. Thus, not only can better utilisation of O_2 under lower ventilation be achieved, but it is also possible to expect a specific effect on the excretion of CO_2. The important point about the lengthening or extension of the expiration mentioned here, however, is that it lowers the excitability of the organism[8] and the relaxation is thus more intense (except where breathing exercises involve forceful expiration), complete full relaxation only being possible during and after expiration.

Besides the 'negative induction' which can lower the level of excitability, there is another mechanism for decreasing the excita-bility of the structure in question. Systematic stimulation of the mucosa and reflexogenic areas leads to their habituation so that the threshold becomes higher and the rate of unnecessary reac-tions to natural stimuli diminishes. This aspect of yoga can be very useful for modern man. His primitive 'fight/flight' mechan-isms, originally needed to deal with threatening situations, are now no longer as necessary in a society where conflicts are largely resolved verbally. The energy resources that are mobilised in the blood stream and subsequently unused by this primitive response are stored in the wall of the blood vessels and can contribute to atherosclerosis.[9] Yoga can be useful here in teaching people to control their reactions to such stimuli, thus reducing the risk of cardiovascular disease.

EXCITATION

The opposite types of exercise, i.e. having a stimulatory effect, include some of the bandhas, kriyas and pranayama. Performed cautiously, according to yogic tradition[10] and the student's capacity, they may not have a particularly significant effect. However when the student reaches the advanced stage and can perform them in a more demanding way[11] this results in a verifiable change in homoeostasis.[12]

Kapalabhati with a respiratory rate of 2.4 to 3.0 per second results in a significant increase in the heart rate of about 30 per cent. Just as we train our skeletal muscles by means of systematic weight training, similarly by systematic stimulation of the organs, the regulatory processes can be stabilised.

Nauli influences the excitation and inhibition of the central nervous system. It produces a complex EEG pattern consisting of large amplitude sharp waves of 12 to 22 cycles per second and a mu-rhythm of higher frequency than normal (11 to 17 cycles per second), an effect which is not inhibited by movement. This has been described as a new rhythm, called xi.[13,14] Similarly, kapalabhati produces a frequency shift towards a slightly increased EEG frequency in the range of alpha band, together with a power increase[15] in the electrical output of the central nervous system, as measured by EEG.

EEG changes during bhastrika were found to be more pronounced than in kapalabhati in terms of the excitatory effect.[16] This depends on the influence of the respiratory centre and upon rhythmic stimulation of the viscera. During bhastrika, the kapalabhati component stimulates the abdominal viscera and plexuses rhythmically, and such rhythmic stimulation greatly influences the functional tone of the central nervous system.

There is an impressive similarity between our observations of ectopic heart activity during nauli and cardiometric changes in a rabbit given electrical stimulation of the hypothalamus in an experiment described by Ulyaninski, Stepanyan and Krymski in 1980.[17] We therefore suggest that, apart from the stimulation of the heart in particular hatha yoga exercises by direct mechanical pressure, most effects are realised indirectly through the central nervous system.

There is another possible excitatory mechanism responsible for the effect of some hatha yogic practices, for example in kriya; stimulation of one structure (e.g. digestive canal) can lower the excitability of another structure (e.g. respiratory pathways). This phenomenon is called negative induction (in the Sherringtonian-Pavlovian sense). In this way we can understand, for example, the acute effect of kriya (when compared to other yoga exercises) upon the bronchial asthmatic. Bhole observed in 1975[18] that asthmatic patients were able to overcome actual or anticipated asthmatic attacks by means of vamana dhauti. Udupa also obtained encouraging results with kriya, pranayama and asana in bronchial asthma[19], and Sedivy described a positive effect of sutra neti in the prevention of migraine.[20] From these experiments it is apparent that kriya, which does not receive as much experimental attention as some other practices, has much to offer in a therapeutic sense.

AWARENESS

Thus we have shown both excitatory and inhibitory mechanisms in hatha yoga exercises and we can accept that yogic methods introduce very marked changes in distribution of excitation and inhibition, both in the brain and in the whole organism.

A further important mechanism influenced by yoga practices is consciousness itself. The restriction of consciousness induced by certain practices such as mantra meditation[21] increase suggestibility, and thus the possibility of using autosuggestive influences to control and improve medical conditions may have great therapeutic potential. Investigations by Wallace, Benson and Wilson in 1971 accord with experiments in attention carried out by Chmelar, Kras and myself in 1976[22] which show that monotonous and long-lasting stimuli lead to a restriction of consciousness within a limited part of the nervous system, whereas other parts are inhibited. This was also the case in EEG experiments with meditating subjects.[23] Stimuli producing alpha blockade in EEG in a relaxed state became ineffective, as a consequence of restricted excitability of the central nervous system.

This corresponds to what used to be called 'watching points', used to describe, for example, the phenomenon of the sleeping mother with a suckling infant who responds to the slightest sound her child makes but cannot be roused by much louder noises. This is an example of a restricted system with a higher level of excitability working in the framework of the inhibited brain. It should be noted here that in the case of meditation, different types and intensity of meditation can evoke different results.[24]

CONDITIONAL REFLEXES

One of the most fundamental physiological mechanisms explaining the influence of yoga exercises upon the human organism is the establishing of conditional reflexes. Especially in those cases in which the involuntary system comes under the influence of the verbal (i.e. voluntary) faculty, conditional reflexes play a decisive role. For example, vamana dhauti produces voluntary vomiting; in the first stage it is induced voluntarily by means of an unconditioned reflex – the unconditional stimulus being the touching of the root of the tongue (and eventually the nasopharynx as well) by the fingers. The conditional stimulus here is the imagination of the procedure and a particular posture. With repetition, the imagination and the posture are sufficient to produce vomiting as a conditioned reaction to conditioned stimuli.

CONCLUSION

Thus we can conclude that hatha yoga is an elaborate system which can be used for improving homoeostasis and, if used rationally, no undesirable side effects should be expected. Hatha yoga procedures are graded according to difficulty, so the approach begins with the easy practices (suitable for patients) and progresses to advanced ones, according to the capacity of the student. The system includes both relaxing (especially suitable for therapy) and stimulating (effective for prevention) procedures. The experimental work done in this field, some of which we have mentioned here, indicates that various mechanisms are at work in the practices, correcting imbalances, and that there are many conditions to which these practices could be successfully applied. Further, in some conditions, yoga therapy can provide a useful adjunct to or support for the maintenance of stability as, for example, in orthopaedic practice. Perhaps one of the most important applications of yoga therapy is in conditions where dependence on drugs can be diminished or removed altogether and the patient's own healing mechanisms can be strengthened, so enabling him to cope with the conditions of modern civilisation in a truly healthy way.

REFERENCES

1 K. K. Datey, S. N. Deshmukh, C. P. Dalvi and S. L. Vinekar, 'Shavasan: A yogic exercise in the management of hypertension', *Angiology*, 20, 1969, pp. 325–33.

2 K. N. Udupa, *Disorders of Stress and Their Management by Yoga*, Benares Hindu University Press, Varanasi, 1978.

3 P. V. Karambelkar, M. V. Bhole and M. L. Gharote, 'Muscle activity in some asanas', *Yoga Mimamsa*, 12, 1, 1969, pp. 1–13.

4 C. Dostalek and V. Lepicovska, 'Hathayoga – a method for prevention of cardiovascular diseases', *Activ. Nerv. Sup.*, 24, suppl. 3, 1982, pp. 444–52.

5 T. H. Schmidt and T. M. Dembrowski, 'Cardiovascular reactions and cardiovascular risk', in *Biobehavioral Bases of Coronary Heart Disease*, S. Karger (W. Germany), 1983, pp. 130–74.

6 *Ibid.*

7 J. S. Meyer and F. Gotoh, 'Metabolic and electroencephalographic effects of hyperventilation', *Arch. Neurol.* 3, 1960, pp. 539–52.

8 A. I. Roitbak, 'Variation of reaction time as a function of respiration phase; possible causes of this phenomenon' (Russian), *Studii si Cercetari de Neurologie*, 5, 4, 1960, pp. 559–65.

9 J. Charvat, P. Dell and B. Folkow, 'Mental factors and cardiovascular disease', *Cardiologia*, 44, 1964, pp. 124–41.

10 Swami Kuvalayananda and S. L. Vinekar, *Yogic Therapy*, Ministry of Health, Govt. of India, New Delhi, 1963.

11 P. V. Karambelkar and M. V. Bhole, 'Heart control and yoga practices', *Darshana International*, 20, 2, 1971, pp. 63–9.

12 C. Dostalek, 'Some aspects of yogic exercises and physiology', in S. Digambarji (ed.), *Collected Papers on Yoga*, Kaivalyadhama, Lonavla, 1975, pp. 33–6.

13 C. Dostalek, E. Roldan and V. Lepicovska, 'EEG changes in the course of hathayogic exercises intended for meditation', *Activ. Nerv. Sup.*, 22, 2, 1980, pp. 123–4.

14   C. Dostalek, M. L. Gharote and E. Roldan, 'Agnisara and xi-rhythm in the EEG', *Yoga Mimamsa*, 22, 384, 1984, pp. 45–50.

15   E. Roldan, Z. Bohdanecky, C. Dostaleck, P. Lansky, J. Los, M. Indra, V. Lepicovska and T. Radil, 'Computer analysis of EEG during some hathayogic exercises', *Activ. Nerv. Sup.* (Praha) 23, 1, 1981, p. 49.

16   C. Dostalek, E. Roldan and V. Lepicovska, 'EEG changes in the course of hathayogic exercises intended for meditation', *Activ. Nerv. Sup.*, 22, 2, 1980, pp. 123–4.

17   L. S. Ulyaninski, E. P. Stepanyan and L. D. Krymski, 'Cardiac arrythmia of hypothalamic origin in sudden death', in *Sudden Death, Transactions: 1st Soviet-American Symp.*, Moscow, 1980, pp. 368–79.

18   M. V. Bhole, 'Rationale of treatment and rehabilitation of asthmatics by yogic method', in S. Digambarji (ed.), *Collected Papers on Yoga*, Kaivalyadhama, Lonavla, 1975, pp. 105–14.

19   K. N. Udupa, *Disorders of Stress and Their Management by Yoga*, Benares Hindu University Press, Varanasi, 1978.

20   J. Sedivy, *Yoga As Seen By A Physician* (Czech), Olomouc, Prague, 1979.

21   R. K. Wallace, H. Benson and A. F. Wilson, 'A wakeful hypometabolic physiologic state', *Amer. J. Phys.*, 221, 1971, pp. 795–9.

22   V. Chmelar, C. Dostalek and H. Kras, 'Relationship of vegetative activity and EEG to active acoustic attention' (Czech), *Cs. Psych.*, 20, 1976, pp. 217–20.

23   B. K. Anand, G. S. Chhina and B. Singh, 'Some aspects of electroencephalographic studies in yogis', *EEG Clin. Neurophysio.*, 13, 1961, pp. 452–6.

24   C. Dostalek, 'Yoga: a returning constituent of medical sciences', *J. Res. Edu. in Med.*, 1, no. 3, 1982, pp. 9–15.

PART II

The Practices

5 The Teacher's Role in the Therapeutic Process

MAUREEN LOCKHART

Therapy is only another word for *help*. Yoga therapy is the art of helping people to help themselves become whole, healthy or well. Mostly, yoga is taught in classes by teachers who may be qualified to teach but who are not necessarily therapists. Some people in the yoga field might say that there is no difference between these roles, since yoga is wholistic and, therefore, therapeutic, but that depends on how it is taught. There are as many different types of yoga teachers as there are approaches to yoga, the spectrum ranging from those who are born healers to physical exercise instructors whose methods are closer to military training than spiritual growth.

In the latter group, the question of therapy does not arise. A teacher who is merely concerned with the physical body, ignoring the emotional, psychical and spiritual aspects of his or her students, is unlikely to have evolved to a level of awareness where a therapeutic relationship is possible. A teacher who really wants to help students become whole, or integrated, must have an integrated view and teach an integrated method and this requires a certain level of development and personal maturity.

Most teachers are motivated to teach because yoga has helped them. The need to help others is not only what Jungian psycho-analyst Irene Claremont de Castillejo calls 'The most human of human needs'[1], but is itself a sign of growth and an important step in personal development. However, not everyone who has been made fit by yoga is fit to pass it on. It is not enough to have had an experience. One must be able to see it clearly enough to communicate it to others so they can experience it as a *whole*.

It doesn't take long for the newly qualified teacher to find out that students don't just sit there and 'do' yoga. They come and tell you their problems. What's more, they tell you about their friends' and relatives' problems – and they expect you to provide solutions as if you had some universal panacea to cure everyone's ills. This can be rather overwhelming for the inexperienced teacher and make her doubt her ability to cope, let alone help in any constructive, therapeutic way. Her enthusiasm and limited repertoire of techniques might seem poor tools to deal with the vast array of problems that her students unburden on her. She knows that the yoga process *is* therapeutic, that time and application *will* heal, but how can she be sure that her method of teaching can help everyone when individual needs seem to be so different? How can she best fulfil her role as teacher in a truly therapeutic way? How can she even be sure that her students will survive her teaching today and come back next week?

No one can really teach another how to teach, any more than to write, paint or compose music. Teaching is a creative art and one only really learns to teach by teaching for, paradoxical though that may sound, experience is the greatest teacher. One can learn a certain amount in training and from observing other teachers at work, but it is from the interaction with students that one really learns.

Teaching is a perpetual experiment, a process of continuous learning in which knowledge grows by being passed back and forth between teacher and student. In this binary relationship it is often difficult to tell just who is teaching whom. Many teachers have remarked that students learn from watching teachers teach themselves; and the teacher in turn learns what it is that the student does with the teacher's suggestions, derived from cues and signals picked up from the student, and how to guide him naturally to the next step.

In this respect the teacher has something in common with the therapist, for, according to Swami Ajaya, who is both a clinical psychologist and a yoga teacher:

The content of each psychotherapy session originates primarily from the client: the therapist is responsive to the client's immediate needs and issues. The therapist does not usually respond with didactic teaching, although he may do so occasionally at the request of the client . . . as is appropriate to the symptoms or issues raised by the client.[2]

The yoga teacher also has in common with some therapists certain advantages that are useful in a directly therapeutic sense. First, she is not teaching a subject from a textbook but imparting something from her own life and experience, so she has a store of *prior* knowledge to draw from that can be of *practical* use in helping others. Secondly, she is likely to have that most important prerequisite for all therapy – *empathy*. This is especially the case if she was drawn to yoga because she has suffered herself, for 'a case might be made out for the view that only he who suffers can be the guide and healer of the suffering'.[3]

The reason why it is important to realise these assets is that for a class of new students, *you* are yoga. Their concept of yoga is not only what you describe and demonstrate, but what you *are*. You are the model of the goal they hope to attain. They would like to radiate good health and self-confidence; they would like to be supple and flexible; they would like to be calm and have their lives under control. And they would also like to be convinced that this desirable state of being that you have attained is not beyond their reach. So they need to be able to identify with you and relate to you. They need to be able to trust you and talk to you freely. In fact, their relationship with you is what the therapeutic process hangs on. I would even go as far as to say that, in the beginning, the relationship between student and teacher *is* the therapeutic process.

Most yoga students are already open to the idea that self-help is the best kind. Just being there in class is the first step towards helping themselves. It also indicates that, no matter how much their health or their ability to cope may have deteriorated, they still have the *will* to survive, for their most important survival mechanism is still intact.

Some of us are better survivors than others. Why is this so? Is it because our parents are good survivors? Obviously not. There are too many examples of children who have survived an early upbringing with sick, unhappy or disadvantaged parents and have grown up to be healthy, happy and successful. What, then, does one require to be a good survivor?

We need to be able to be self-sufficient and adaptable, able to handle the challenges and the opportunities that life presents us with. But our lifestyles may be so demanding or distressing at times that none of our available skills are of any use to us. We do not have the knowledge, the experience, or even the energy, to cope. What then? How do we survive crises and grow through them?

We seek the help of others. And this may be the most vital skill we possess. To be able to ask for help in a vulnerable situation is not a sign of weakness but of emerging strength, because it is a natural, in-built survival mechanism, evolved over centuries.

Several studies of the relationship between emotions and health show that a person's self-esteem and coping capacity is enormously dependent on his ability to form bonds with others and that this, in turn, affects his health. In one study conducted in California with 7,000 adults, it was found that 'men and women with firm bonds of family, friends and church had less than half the mortality of those without the comfort and balance of such ties'.[4] And it is now widely recognised, largely due to the work of stress pioneer Hans Selye, that bottled-up emotions – feelings not shared with others – severely challenge our survival response and lead to migraines, hypertension, heart disease and a host of other stress-related illnesses. 'Attachment bonds,' says Dr Gerald Klerman, 'have been useful, even essential to the survival and development of our species. They are served by a psychobiological apparatus developed through centuries that we have inherited from our mammalian ancestors.'

When we are able to reach out without fear for a helping hand and to let go when we are able to stand alone again, we have taken our first step towards growing up. From infancy to old age, this pattern repeats itself with every new experience. This is how we survive. And our need to survive is so powerful an instinct that, each time we encounter changing circumstances that we cannot meet entirely through our own resources, nature evokes the need for reinforcement. We 'yoke' our 'selves' to an 'other' to help us get through.

Yoga means yoke, from the Sanskrit word yuj, and I believe that the relationship between yoga teacher and student is an important, temporary bond formed for the purpose of survival. It is a unique relationship that has grown out of a change in our social structure, a need created by the prevailing ethos of our society – our separatist way of thinking – for it seems to me that the real disease of the twentieth century is not stress, which is merely a symptom, but *separation*.

> The tragedy of today is that civilisation in the West has gone too far in the direction of masculine separation and the masculine urge to discover and create for its own sake and the accompanying danger that women are throwing over their sense of cosmic awareness and the connection of all growing things in order to adopt men's values in their stead. That only serves to upset the necessary balance between the opposite poles.[5]

We are, as philosopher Richard Rorty says:

> the heirs of three hundred years of rhetoric about the import- ance of distinguishing sharply between science and religion, science and politics, science and philosophy, and so on. This rhetoric has formed the nature of Europe and made the West what it is today.[6]

And what of our natures? What has this done to us as human beings? The legacy we have inherited is the separation of our- selves from nature and of one part of our nature from another – intellect from intuition, thinking from feeling, mind from body – with the result that it is hard for us to survive as whole human beings, or even to know what wholeness feels like. And yet we do seem to have retained that particular memory, for we are older than a mere 300 years and there is some evidence that this state of separation or dis-unity is something we are taught, or con- ditioned, to believe, for 'Most children are born with, and many children retain, a diffuse awareness of the wholeness of nature, where everything is linked with everything else and they feel themselves to be part of the individual whole.'[7]

It is not surprising, then, that if our attempts to survive as a whole in a society that compartmentalises us make us sick, and that if we can no longer count on traditional bonds with doctors, priests and professional helpers for comfort, reassurance or even attention, then we find ourselves looking to conditions in other cultures for our salvation. For we have a biological imperative to survive, no matter what. The evolutionary journey must go on and if obstacles to our survival appear which we cannot remove, then it is our natural instinct to find alternative pathways and alternative vehicles or methods to help us get through. Hence the growing popularity of yoga, whose ethos and practices provide the opportunity for healing those disparate elements of ourselves. Furthermore, as Larry Dossey points out in *Space, Time and Medicine*:

> almost by definition health is itself a survival strategy. It is practically axiomatic that sickly members of a species would be relatively unsuccessful in perpetuating their genetic structure. Being sick is a poor survival strategy, just so is isolation. If both being healthy and being in association with one's kind confer survival advantage, we might expect to see these qualities coupled in individuals.

Knowingly, or unknowingly, every person who seeks out a yoga class is really looking for a therapeutic environment in which to become whole. Though they may express this need in many different ways — as medical symptoms, emotional problems, feelings of loneliness or inadequacy, or the need to explore and expand personal development — each person is looking for a way of learning the art of survival so that they may become fit to survive. The yoga teacher is the mediator and the guide in this process, with whom the student forms a temporary attachment bond in order to grow and evolve into a self-actualised human being, a survivor.

Through the medium of this attachment, which is based on mutual trust, respect and sharing, the individual's sense of separation gradually disappears and a feeling of belonging emerges. This feeling is reinforced by the support of others in the group as well as a growing awareness of his relationship to life and the cosmos at large. When this sense of belonging really 'takes hold', he stops being separated and imprisoned by his ego; he is able to 'let go', detach himself from the teacher and become independent and self-sufficient.

There are three major steps in this process that the teacher needs to know, and he must know how and when to initiate them.

1 The first is to provide a healing environment.
2 The second is to have a therapeutic method to lay down principles for the stages of revitalising, cleansing and healing.

3   The third is to provide guidance in personal exploration through meditation, discussion and counselling.

A HEALING ENVIRONMENT

This is largely created by the teacher's attitude and personality. A person in need of help is more likely to be receptive to someone who is gentle, encouraging and understanding than to a task-master who is only concerned about correct performance. People suffering from distress – especially those students for whom yoga is the last resort in a long line of frustrating attempts to find help for a chronic condition – need to be able to let go of anxieties in a safe place with a person they can trust. They must be sure they will not be humiliated or disapproved of. A class full of consenting adults should not be treated like naughty children. Failing to get something right is not disobedience and does not require punishment. On the contrary, making mistakes is a valuable means of learning for both student and teacher. 'Misfortunes and mistakes,' says Irene Claremont de Castillejo, 'are part of our pattern. Even illnesses may be part of our pattern to give us pause or teach us lessons we should not otherwise learn.'[8]

Clear directions and tactful suggestions, drawing a student's attention to particular details, are more effective in the long run than placing a person 'in position'. The freedom to ask questions about the work in hand and to have an unlimited number of demonstrations or explanations will draw from the student his *own* best performance according to his level of awareness, which will improve from session to session. Furthermore, a policy of 'non-interference' allows the teacher to monitor the student's development more closely; it gives necessary feedback about the student's ability to handle and utilise new information, to relate it to existing knowledge and to apply what he has learned by observation and experiment. Growing inner confidence and a balanced state of mind shows outwardly in improved coordination and control, in a change of posture and demeanour and, ultimately, in a change of attitude even towards the process itself. Students who were apathetic become more positive and enthusiastic; latecomers become more punctual; most students begin to integrate their practice into busy schedules and accept the value of yoga as an investment in their own health.

The most important factor in creating the right kind of atmosphere within the class is *enjoyment*. When fear of perfect performance is removed, the student is able to relax and let go of tensions more easily; it increases his confidence and allows him to be an individual within the group, developing at a pace that is appropriate for him instead of worrying about 'keeping up'. He is then able to enjoy the process and let the goal take care of itself, so paving the way for healing to begin. It also has another valuable therapeutic effect; it helps the student let go of feelings of anger or helplessness brought on by his condition. As Dr Marjorie Brooks puts it:

> Often people feel angry over being dependent and helpless in a treatment situation. But if that angry energy can be redirected, they feel less stress, and quite likely their physical condition will be positively affected . . . activities that involve physical motion are good for discharging angry feelings.[9]

This is also the case with people who may feel angry or helpless in a personal or working relationship. Participation in a self-help method that allows them to restore confidence and self-esteem is an important step in enabling them to see their situation more clearly and to resolve it through a change of attitude and/or appropriate action.

Teaching requires an intense level of concentration to produce a high level of energy, for although healing comes in many forms, a high level of vitality is fundamental to all of them. Being in the presence of a teacher who emits 'good vibrations', with whom they feel at ease, is beneficial for students and helps them to raise their own energy level. In the beginning students may be dependent on sharing the teacher's energy. Until they are fully 'awakened', students are often 'sleeping partners' in the therapeutic business. They are often preoccupied with an energy-squandering attachment to their 'circumstances' or 'condition'. When they decide to let go and stop attaching blame to past events they are able to concentrate fully on the present and surrender to the process. At this point they begin to improve as they become self-motivating and generate their own energy.

In his book *The Global Brain*, Peter Russell explores this experience of sharing energy – common to teachers, therapists and healers – by considering a parallel phenomenon in the world of physics, the functioning of a laser, through which he explains the effect that a small number of 'enlightened' people (that is, people actively working on inner development through consciousness-raising techniques) could have on the rest of society.

> Light from any source consists of numerous different tiny packets of waves (quanta), each coming from a different atom. In ordinary light these waves are generally all out of step; they are said to be 'out of phase'. If, however, during the brief instant that an atom is about to emit its minute wave packet, light of a specific frequency (or colour) impinges on it, the atom can be stimulated to emit a light pulse that is in phase with the wave that stimulated it. The new emission thus augments, or amplifies, the passing wave. At low power the net effect is still one of bundles of waves, out of phase with each other. As the power is increased, a certain level, called the laser threshold, is reached at which a completely new phenomenon occurs: all the little bundles suddenly lock into phase; they are said to become coherent. When they do so, there is a tremendous increase in the intensity of the light produced . . . thus a small number of units acting coherently can easily outshine a much greater number acting incoherently.[10]

Being absorbed by the teaching process is itself a healing, energy-raising experience, for it allows one to forget oneself and to become one with – yoked or joined to – the yoga process. Sharing energy means encompassing someone else in our world and breaking down the barrier of ego separation. The more often we do this, the more aware we become that we all belong to the whole, we are all *one* energy. Sharing your energy with your students gives them the impetus to grow. They are able to go beyond the limitations of their individual identity and merge with the whole. Helping students to raise their energy levels through yoga techniques filtered through your own experience and imparted with enjoyment is a healing experience. When students leave a class feeling relaxed, more energetic and better equipped to cope with their lives, the experience has been truly therapeutic. They are well on their way to identifying their own problems and becoming their own therapists, but in order to develop the necessary skills they need to be presented with a coherent method.

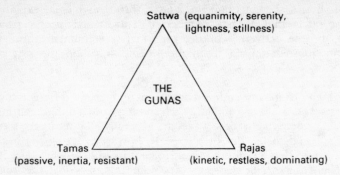

Figure 5.1   The triangle, symbol of the gunas, the three energy states that are in constant interplay, makes a useful paradigm or teaching model to help students understand the relationship of yoga practices.

A THERAPEUTIC METHOD

This is essential in order to ensure that the technique, the pace and the application is right for each person at every stage. This is not an easy task, for one of the most disturbing factors about class work is that while some students are constant in their attendance, others come and go, with the result that different students are at different levels in their understanding of yoga. To some extent this is to be expected, since their individual capacities, their speed of learning, their age, their state of health, all influence their response to the yoga process. In such circumstances I feel that effective therapeutic work can only be done in small groups. With a maximum of ten students in a class, the teacher is in a better position to observe each person and have a deeper understanding of individual needs. I also find it practical to advise new students to think in terms of committing themselves to three months (12 weekly sessions) in order to establish a solid foundation in their understanding of yoga practice.

In order to help students understand the complexities of the many techniques, their relationship to each other and the whole, I find it useful to use a paradigm, or what one might call a 'teaching model'. Experimentation has shown the triangle, symbol of the gunas, the three states of energy (see Figure 5.1), to be most effective in helping beginners understand the underlying pattern

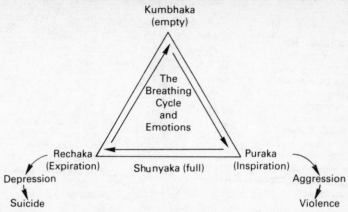

Figure 5.2 The natural motion of the breathing cycle can be disturbed by emotional reactions to stressful or threatening situations (real or imaginary) fixing the movement in one direction or another and creating an imbalance that affects behaviour.

of the yoga process. Students are better able to comprehend concepts like energy, balance and integration when they have a visual model to relate to and through which they can provide themselves with feedback to monitor their own progress. Through this model, students learn that the objective of yoga practice is to bring about the dynamic energy state of sattwa guna which gives a feeling of lightness, weightlessness and serenity, a perfect state of alignment in which there is also stability and strength. The focal point of yoga practice is the breath, our primary source of prana (energy or life force), the bridge between mind and body that provides a kind of two-way mirror in which changes of balance are reflected. Students thus learn to observe how disturbances in the rhythm of the breathing pattern created by thoughts can alter emotions and moods (Figure 5.2), which in turn show up in the physical body as muscle tensions or loss of tone which alter posture and alignment (Figure 5.3).

Through this simple teaching model students learn how to utilise easy breathing exercises to restore natural breathing and harmonious balance. They learn that since the natural motion of the breathing process is cyclical – the in-breath lifts the pelvis up and pushes the chest forward, while the out-breath does the

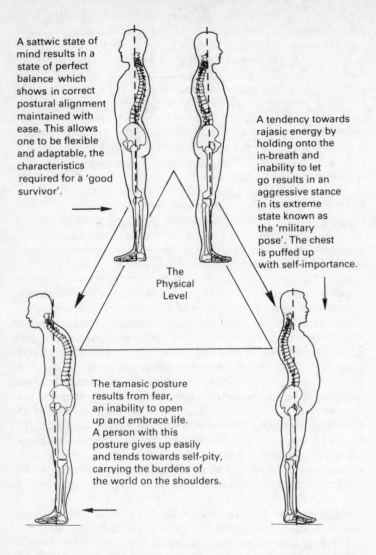

A sattwic state of mind results in a state of perfect balance which shows in correct postural alignment maintained with ease. This allows one to be flexible and adaptable, the characteristics required for a 'good survivor'.

A tendency towards rajasic energy by holding onto the in-breath and inability to let go results in an aggressive stance in its extreme state known as the 'military pose'. The chest is puffed up with self-importance.

The Physical Level

The tamasic posture results from fear, an inability to open up and embrace life. A person with this posture gives up easily and tends towards self-pity, carrying the burdens of the world on the shoulders.

Figure 5.3 Alterations in the breathing cycle created by particular thought patterns ultimately manifest themselves in a change of body shape. Yoga practices, especially asanas, correct alignment and restore harmony to emotions and to the mind.

opposite, making the chest sink, the back round and the head drop slightly – a disturbance in this movement, fixing or 'freezing' the movement in one direction or another, creates an imbalance. Disturbances (kleshas) are fixed attitudes of mind or the habit of reacting emotionally in a particular way to stressful or threatening situations (real or imaginary) which alter the breathing pattern and which in turn interferes with innervation, endocrine activity, digestion, circulation and numerous other body functions and produces chronic muscular tensions. Muscle spasticity then alters the skeletal configuration, creating rigidity and loss of freedom of movement.

Observation of the breath provides the student with the key to his recovery and the teacher with an important diagnostic tool. Through observation of the breathing pattern, the teacher can teach the student to correct a tendency towards rajasic or tamasic behaviour for people tend to fall into one of two categories – 'in-breathers' and 'out-breathers'. 'In-breathers' can hold on to the life force but cannot let go; 'out-breathers' can give up but cannot hold fast. Most people have an imbalance to one degree or another that shows up in their attitudes and in subtle or gross alterations in vertical alignment of the spine and the horizontal planes. In effect, every body has a different shape that reveals a pattern – the result of attachments to thoughts, feelings and habits. This pattern reveals what we might call our 'stepping style' – the way we relate to the world, the way we handle pain and pleasure. Our body shape contains both our history and our future, our inner and outer selves.

It is the body, as a rule, which flourishes exceedingly, which draws everything to itself, which usurps the predominant place and lives repulsively emancipated from the soul. A human being who is first of all an invalid is all *body*; therein lies his inhumanity and his debasement.[11]

Concentration on the breath frees the person suffering from an acute or chronic condition from the tyranny of body consciousness and raises his level of awareness to other aspects of the self by teaching him the corrective or curative principle on which all yoga techniques are founded, that *every situation must be taken to its full conclusion in one direction or another so that balance, or 'harmonious vibration' can be restored.* Furthermore, the health of every cell depends on its vibratory activity and this, in turn, on the vital force or prana flowing through it. Lowered vital force means lowered resistance and a weakened ability to expel toxic matter from the body, preventing nutrition from reaching the cell, which further reduces vitality.

Starting every yoga session with instruction on the natural way to breathe, the performance of the complete breath and other simple breathing exercises, not only raises the level of vital force through the in-breath, but expels unwanted toxins through the out-breath, thus revitalising and cleansing – the first two steps essential to the healing process. Natural breathing relieves tension in the muscles, releasing energy. It teaches the student how to conserve energy by focusing the mind, thus preventing the squandering of energy on uncontrolled compulsive thoughts, which in turn helps to control emotional reactions and to restore equilibrium in stressful situations.

This prepares the student for step 2, in which simple stretching movements, coordinated with the natural movement of the breathing rhythm, are introduced without disturbing concentration. This teaches the student how to conserve energy in movement by keeping the mind still while the body is in motion. Through these stretching movements, he learns how to relate to gravity and corrects his symmetry and polarity, which shows up in improved balance and coordination. Having understood the relationship of breathing to movement, of stillness to motion, holding on to letting go, the student is better able to 'stretch from pole to pole in order to achieve the paradox of holding both attitudes at once'[12] and to move gracefully and easily in and out of postures with the minimum loss of energy and self control.

Postures give stability, strength, endurance, so increasing recuperative power. They help the student to recover sensory awareness, sharpening his perception about potential health-threatening changes in his metabolism and alignment. Flexibility is increased through the range of positions – forward, backward, sideways, laterally and inverted – not only toning muscles and improving posture but also increasing his pain threshold and his confidence in relating to changing situations, helping him to meet new challenges and opportunities for growth beyond his self-limitations and thus invoking his own healing power.

The paradigm or teaching model that provides the pattern for these three steps – breathing, breathing plus movement, and breathing plus movement plus posture – is equally effective in providing a sense of continuity throughout the more advanced practices of pranayama (breath control), pratyahara (sensory withdrawal), dharana (concentration) and dhyana (meditation), for it is a constant reminder of the three qualities of attention – *balancing*, *directing* and *focusing*, or equipoise, motion and stillness – reflected in the natural rhythm of the breathing cycle which mirrors the pattern of life itself. Systematic application of these three qualities to the higher stages of practice helps to reduce confusion, conserving energy and encouraging students to practise more consistently at home. For example, I find that students more readily apply themselves to meditation when they learn first to practise each of the previous stages in three steps. For example, pratyahara can be learned by:

1 balancing energy by extending the in-breath upwards and the out-breath downwards, gradually, bit by bit, breath by breath, until students are breathing from the crown of the head to the soles of the feet, so encompassing the whole of the body in conscious awareness;
2 directing energy through shavayatra (inner travelling); that is, breathing into specific areas of tension like the joints and/or muscles;
3 letting the mind come to rest in a specific place, such as the ajna (brow) chakra while observing the ebb and flow of the breath with passive concentration, thus preparing the student for the next stage of dharana (concentration proper) which can be similarly introduced in three steps.

Experiments with various techniques of meditation in similar fashion allows meditation to be introduced as a natural extension of the previous stages of practice, so that students do not develop notions that meditation requires a different attitude or approach but are grounded in the knowlege that all practices adhere to the same basic principles.

Students also find it easier to understand the philosophical premise of yoga (Figure 5.4) through this paradigm when they see that the triangle (symbol of the gunas) represent the three states of consciousness – unity, duality and diversity (Figure 5.5). For example, the new-born baby has unitary consciousness until it develops a separate identity or ego – the 'I'. As it learns to relate

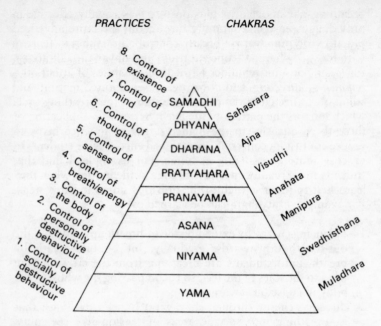

Figure 5.4 The classical yoga model, Patanjali's eightfold path, gives the impression that development is linear.

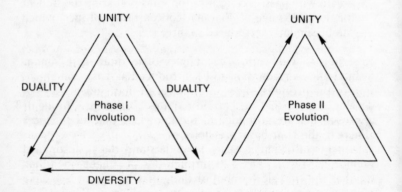

Figure 5.5 The gunas can be used to represent three phases of consciousness that are experienced from birth.

to the world (phase I) and separates itself from its mother, it has to develop numerous or diverse personalities to survive. In the second phase of life (phase II) we are less concerned about making our mark on the world; we are more concerned with developing our inner rather than our outer resources. We evolve a more singular or unified 'self' as we grow towards death and reunification with the source of life (however we may express it).

When the students become aware of how they identify with one polarity or another of the triangle, they learn to transcend these states and can direct their energies from a higher level of awareness and a new perspective, in which they can see their relationship to the whole and relationship of the parts, through yoga practice, to each other.

This model may also express the brain, with its two hemi-spheres, or the masculine and feminine aspects of personality, which can be united through meditation. Control over the mind results in the cessation of disturbances, the toing and froing of energy from rajasic to tamasic, and vice versa, and brings stillness and peace through one-pointed concentration.

The yoga process teaches us to attain mastery over our physical, mental/emotional and spiritual bodies, enabling us to expand and explore our potential. This can give us freedom from physical illness, mental tensions and emotional conflict and take us into the realms of experience as yet not fully explored by orthodox science.

Thus the teaching model can be used to show that the classical yoga model, Patanjali's eightfold path illustrated in Figure 5.4, far from being a linear development, in fact consists of a number of overlapping steps (Figure 5.6). Each step consists of a whole group of practices designed to help the individual evolve into a more aware being and free him from his self-limitations. The first four stages of the classical model (Figure 5.7) are traditionally represented as upward pointing triangles, symbolising the desire for unification of the material (earth, water, fire, air) with the spiritual. Mastery over these levels leads to expansion of the higher chakras (represented by the reversed triangles) through mind to spirit. The first and seventh chakras and/or steps are furthest apart and the fourth (prana) and fifth (senses) closest together, these last two being the link that provides the bridge between body and mind.

Figure 5.6 The teaching model can also be used to show that Patan-jali's eightfold path in fact consists of a number of overlapping steps.

Taking the teaching model even further, the Sri Yantra (Figure 5.8) is often used as a concentration symbol to help aspirants liberate themselves from the limitations of form (material level). As they evolve, expansion is not only vertical but also horizontal; gradually awareness extends in all directions, enabling the student to understand his relationship with all manifested forms of life. In practice, this represents polarising and realignment of the body so that energy may flow through the sushumna (central spinal channel). When this occurs the student has an experience

Figure 5.7 The triangles can be inverted to represent the higher chakras.

Figure 5.8   Triangles combined into the Sri Yantra.

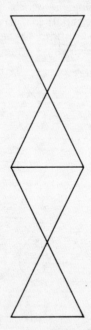

Figure 5.9   Triangles used to represent the rhythm of expansion and contraction.

Figure 5.10 Figure of eight, used to show how the ida and pingala intertwine.

of expanded energy and well-being. As the student gains control of the various stages, so his chakras (energy centres) awaken and his energy begins to flow through the proper channels. The movement of energy is felt as a rhythm of expansion and contraction (Figure 5.9), reflected in the cycle of the breath.

The ida and the pingala, which have been related to the sympathetic and parasympathetic nervous systems, are symbolised as intertwining in a continuous figure of eight (Figure 5.10). As they cross over at each chakra (Figure 5.11), so the vibration

Figure 5.11 The intertwining of the ida and pingala, crossing over at each chakra.

of that centre is said to speed up and work at a higher level, improving the health of each region and ensuring efficient functioning. This is often represented as a wheel which is visualised as spinning at a faster rate (imagine Figure 5.8 spinning like a windmill), accounting for the feeling of light-headedness or en*light*enment or an actual experience of power being generated and producing a dazzling internal light.

Using a recognisable working pattern alongside such a teaching model gives continuity from class to class, providing students with a framework within which they can experiment and discover the techniques that are most effective for their particular needs. And, most important, it helps students to concentrate on the process rather than the goal, freeing them from the tyranny of time that Larry Dossey believes prevents people from functioning in tune with their own natural rhythms and harmonising themselves with those around them, and thus providing students with a positive strategy for survival.

PERSONAL GUIDANCE

This becomes necessary only when people ask for it. Since I encourage students to ask any questions they wish, or make any observations, at a special time set aside for that purpose towards the end of the class, most doubts and misunderstandings are cleared up then. At the times when it becomes necessary to introduce topics on nutrition, health care or deeper aspects of philosophy, this time naturally extends itself as and when people develop an interest in and wish to explore these subjects in greater depth.

As students become a more cohesive group, they tend to reorganise their other activities so that they can have more time together at the end of the class, perhaps having refreshments while an open discussion ensues on anything from books to backache, from diet to driving. I've also observed that the personal folder which each student keeps to collect the drawings and material I distribute suddenly expands to include interesting articles and illustrations on a wide variety of topics relating to yoga and health from magazines and journals which students actively participate in searching out to bring to these discussions.

As people get to know each other better, they are able to talk about their own particular problems and to receive support and advice from others. Here, the role of teacher changes slightly, for he/she is not called upon to instruct, or even to talk, throughout every session. This allows the teacher to study people at closer quarters and observe how they relate to others, for, in the beginning, one only sees students outside the context of their lives. However, it is the nature of the yoga process that eventually students become friends and friends become students and they naturally turn to you – as perhaps one of the few people they know who appears to have got their life in order – if they are worried by a health problem or there is something happening in their personal relationships that they cannot handle.

This is where some teachers part company from others, for some believe their 'job' is over when the class ends and the one-to-one situation belongs in another category called 'counselling' or 'psychotherapy'. This is entirely a matter of personal choice, but I believe that if students do not have the opportunity to talk over personal matters, their progress may be impeded. It is often the yoga process that has allowed the problem to come to the surface, and the problem may be the original cause of their 'imbalance'. To deny the student the opportunity to discharge or deal with the cause of their tensions – to take it to its natural conclusion – seems to me to deny the opportunity for healing and taking a valuable step towards wholeness. Besides, for some people psychotherapy is *not* an option. Either they consider their problem not serious enough to warrant that sort of attention, or their social or financial circumstances prevent them from seeking it.

Here I consider meditation to be my most valuable therapeutic tool. The steady watchfulness that meditation training produces enables me to direct the class proceedings while storing up observations that may be valuable clues to each person's development. And since I have always had the practice, right from the very first lesson I ever gave, of meditating on the class at the first opportunity when I reach home and of carrying the memory of the group, collectively and individually, around in my subconscious throughout the week, I have discovered that 'In the unconscious, time has no meaning and all things are already known.'[13]

In meditation, surprisingly clear images of some people emerge instantly, while others, out of focus at first, gradually come forward as the weeks go by. By replaying scenes of action and interaction, of movements made, sentences spoken, questions asked, doubts expressed, opinions ventured, I get to 'know' my students and 'see' their problems. In meditation I become them; I feel their body shape, speak with their voice, experience their behaviour, follow the course of their pain. By the time they have progressed to the point where they are fully involved in their own recovery – which is the time when they usually come forward for individual help – I have a clear picture of both their visible and 'invisible' energy pattern and the spiritual, emotional and physical causes of their dis-ease.

There is no great mystique about this. It is not an extraordinary ability but a potential that we all possess that is developed merely as a byproduct of yoga. As Thomas Blakeslee puts it in his book, *The Right Brain*:

> In this word-oriented world, it is very easy to overlook the fact that we have many other kinds of thought besides verbal thoughts. Thinking, after all, consists of manipulation and rearrangement of memory images. The ultimate source of the memory images used in thought is our senses. Since vision is the most information rich of our senses, visual thinking is extremely important and powerful. The other senses, however, also produce memory images which can similarly be the basis of thought.[14]

Swami Paramahamsa Satyananda Saraswati of the Bihar School of Yoga expresses a similar thought from the yogic point of view. He says that:

> Western psychologists have not concerned themselves with man's spiritual salvation. They are content with curing mental disorders. In their study of the human mind they stumbled upon the discovery that man is endowed with extra sensory perceptive power, and that it is dormant in all of us . . . Our Yogis all along knew about this power. They could see with eyes closed, walk without moving and hear with ears plugged. They practised the *art of awakening* our super psychic centres of deep meditation.[15]

In private discussion with students, however, I make it quite clear that I may not be able to do any more than give them my undivided attention. If I have suggestions to make, they are only suggestions and the student is not obliged to follow them, or even to agree with them. And since I am quite open about the fact that I have no aspirations to sainthood, that I do not know everything, that I do make mistakes and get things wrong on occasions, the student knows that he can depend on me to give whatever help I am able to give, but that he should in no way relinquish responsibility for his own life.

Sometimes listening is all that is required, along with some gentle hints about seeing things from another angle. At other times it may be necessary to refer students to a medical specialist, to another kind of therapist or to employ any additional skills of one's own, such as massage or acupressure, to help alleviate acute distress. Often people want to discuss the content of a dream or meditation and receive assurance that their own observations are valid and are guiding them in the 'right direction'. Frequently, people want to talk about the changes that yoga has made to their lives and how it has altered their outlook, their relationships and expectations. And on some occasions the private session does develop into something overtly resembling psychotherapy, which worries some teachers who feel that they may not have the appropriate skills or training to help their students effectively.

In my experience this situation never arises by intention but by *attention*. Using the same teaching skills that one develops in a class, plus the uninterrupted opportunity to listen, observe and guide a person in the direction he or she appears to need to go, will naturally lead to exploration of deeper aspects of personality. And those same skills will tell you whether you can continue to guide that person or whether they have a deep-seated problem that requires more specialised help.

In many cases, however, insight and intuition gained through meditation will provide the right means to help people in the right way. For example, an osteopath sent me a patient with a breathing problem. My very first impression of the man was 'barrel-chested, hard outside, soft inside'. None of this meant anything to me until I discovered that the man was a journalist with a hard-boiled exterior but a soft heart who had been the victim of a malicious prank as a small child and had nearly drowned in a barrel of water. For the first session I simply watched him go through some easy breathing and stretching movements and it became obvious that he could not breathe in properly. Breathing in for him meant breathing in water and drowning. Any inducement to increase his in-breath brought a panic response. I could see what his problem was quite clearly, but I didn't know if I was going to be able to help him solve it. For several days I focused on him for a few minutes in my meditation session, but nothing happened. Then, around the fourth or fifth day, he appeared in my meditation holding a beautiful long-stemmed red rose. I hadn't a clue what the meditation meant, even by the time my student arrived for his next session.

Then it came to me. I asked him if he liked flowers and when he said he did, I asked him if he'd noticed how one has to smell a flower. He was puzzled by this question until I explained that *trying* to smell a flower doesn't seem to work; the smell eludes one unless you become totally passive and let the fragrance come to you. I asked him to experience this with a real flower; then to imagine he was smelling a flower. Although difficult at first, he learned to breathe passively through this method, gradually deepening his breath so that he could breathe in without tension or fear. Eventually, I just had to remind him to 'smell a flower' whenever he was tense in order to see him visibly let go and relax.

After some practice in concentration we were able to explore the original trauma. He became the little boy in the barrel, but a little boy who, although he'd had a nasty experience, was still alive and had grown up into a man who was still alive and no longer afraid of being in the barrel. It took some time to get through this block but, when we did, this very courageous, gentle, man stopped carrying his barrel around with him; he lost weight, ate better, slept better and, when he telephoned me one day to tell me he didn't need me any more, *was* better.

It is important to note here that one should listen carefully to one's 'inner guru' and learn not to distort insights by interpretation. Although I often do not know at first what these impressions mean, they invariably provide the key to unlocking a person's psyche and opening them up to examining the situations in which they are making themselves or another person unhappy. It should be made clear, however, that these observations in no way imply a judgment of the person. Sometimes they are visual, sometimes verbal. For example, as I was watching a woman with a back problem perform in class, I noticed that her face was always 'closed'. I had the feeling that everything was pushed 'down in back'. Later, in the middle of a private discussion, I found myself asking her if she often felt like crying but pushed the feeling away. She told me she felt she had to because there was no solution to her problem. She was unhappily married to a very nice man who loved her but couldn't understand her. We were then able to explore her needs, her expectations from her relationship, what she should, realistically, look to other relationships with friends to provide, and what she should provide for herself through some creative outlet.

On another occasion a very elegant middle-aged lady who had been coming for private tuition for months suddenly burst into tears in the middle of a posture. She refused to tell me why, saying I was 'too young to understand'. I was about to ask her how she knew I was incapable of understanding unless she told me what her problem was when I found myself remarking instead that I was surprised to find such an 'upright' lady losing her composure. She stopped crying at once and asked what I meant. I tried to explain to her as tactfully and gently as possible that I always thought of her as 'upright' and rather 'tight-laced', like some Victorian lady. I also felt that it must have taken something deeply wounding to have made her lose control.

Suddenly, after months of being very secretive about her private life, she told me she was a widow, having a relationship with a man after a very long period alone. In a disagreement her partner had made remarks about her being 'old fashioned' and 'moralistic' that had hurt her feelings and conflicted with her own self-image. That casual remark of mine had opened the door to exploring her needs, her relationship, her feelings and values, that had a positive effect in helping her to understand herself better and in helping her relate to others.

Similarly, while I was talking to a man I'd never met on the telephone who wanted to come for tuition to improve his concentration, I had a picture of a foot gripping hard on the floor. When this person came for his first lesson I asked him if he'd ever injured his right foot. He said he couldn't remember having done so. However by the time he'd had a few lessons I was able to identify his basic conflict and his reason for not being able to concentrate; he was a musician who desperately wanted to play and teach full time, but his need for security was making him hold on to a job he disliked – and I identified the problem only when I saw him grip hard into the floor with his right foot. When I called his attention to this habit of gripping on for security, he also remembered that he *had* injured the foot playing football when he was a child. And it was really from this experience that I learned that one may be unlocking more than one layer of experience and that this is all the more reason not to tamper with 'inner messages' until the message applies itself to the 'medium'.

Some people's problems are fairly near the surface while others may have deep-seated blocks that emerge like erupting volcanoes. Here, again, it is necessary to stress the importance of listening carefully to the 'inner guru', for one can be given warning signals that a person may be hostile, manipulative or ego-centred. One should then proceed carefully and/or refer the person to another therapist or terminate the tuition. Some men, for example, who have an undeveloped feminine aspect, can be hostile towards a woman teacher and would be more effectively helped by a man who can guide them into an exploration of their feelings.

When people come for private tuition, it is worthwhile having a preliminary chat about their reasons for wanting to practise yoga and/or meditation. This often reveals aspects of character or circumstances in people's lives that will allow you to gauge whether or not you can help them. Of course one cannot be absolutely sure about some people, like the martial arts teacher who appeared to be 'normal' for the first two weeks, then suddenly became abusive about my ability as a teacher when he was not enlightened in the third lesson. While he was busy ranting at me, I had a flash of insight about the uneasy feeling I'd had about him from the first meeting. I realised I'd been 'back pedalling', literally 'drawing back' from him because I distrusted him. I suddenly realised it was because I felt that his ego had been inflated by his powerful position and he hated being a student because it made him feel vulnerable instead of invincible – he was totally frustrated by my treating him as a beginner in yoga instead of the master he really felt he was.

I stopped his monologue by congratulating him on coming face-to-face with his problem so soon. While he was recovering from that statement, I explained that he would never become a master in yoga until he had the humility to be a student and that in yoga we consider ourselves perpetual students of life. I asked him to think this over and decide whether we could continue working together. He never came back, but he did send a friend for tuition. One could interpret that as an acknowledgment that he'd at least listened and, perhaps, even taken the next step with a change of attitude.

Basically this is all one does; help a person take one small step in their own journey by helping them to come to terms with their own needs, their fitness to survive in a society with which they may be at odds, politically, emotionally, ethically or religiously. It is important, therefore, not to indulge in notions of success or failure but to see oneself historically, as a person who holds out a helping hand, a partner in the process of development.

Teachers, we are told, need to be dedicated. Teaching is not a job, it's a vocation. There may be some truth in that, for I find that teaching is part of my life and my life is part of teaching. Or, to put it another way, I teach what I am and I am what I teach. It is extremely unlikely that you, as teacher, have gone through the whole of your life without facing problems, some of which you may yet be carrying around in your system. However, you also have the advantage of having the experience of yoga to teach you how to solve your problems and to recover your balance and harmony, perhaps even helping you recover from illness. So it is natural that you want to share this experience of wholeness with others.

Some yoga purists say that there is no such thing as yoga therapy; that it is not the purpose of yoga to deal with people's health problems. 'There's nothing in the ancient texts about curing illnesses,' they say, 'and, anyway, yoga is a spiritual discipline, not a method for making people healthy.' All of this is perfectly true, of course, but if the purpose of yoga is spiritual perfection, union or 'wholeness', then surely it must mean wholeness of the *whole* person – body, mind and spirit. And since a lack of wholeness leads to a state of dis-ease, or dis-integration, it seems to me to be perfectly legitimate to use appropriate yogic methods to help oneself or another back towards health. How else can we evolve?

Allopathic medicine has done a great deal in wiping out infectious diseases, but infection is not the only cause of dis-ease. And as the World Health Organisation has stated: 'Perfect health is more than the absense of disease. It is a positive state of well-being.' A well being is one who has passed beyond illness.

And evolving toward perfection must surely mean growing out of those 'misfortunes and mistakes' into a new pattern of behaviour. Changing direction is an arduous and sometimes painful task, and it is difficult to do it alone. Most of us need help, at least, in the beginning. And therapy *is*, after all, only another word for *help*.

REFERENCES

1 Irene Claremont de Castillejo, *Knowing Woman*, Hodder & Stoughton, London, 1973.
2 Swami Ajaya, *Psychotherapy East and West, A Unifying Paradigm*, Himalayan International Institute of Yoga Science and Philosophy, Chicago, 1983.
3 Thomas Mann, *The Magic Mountain*, Penguin, London, 1960.
4 Maggie Scarf, *Unfinished Business*, Fontana, London, 1981.
5 Irene Claremont de Castillejo, *Knowing Woman*, Hodder & Stoughton, London, 1973.
6 Rudolph Otto, *Mysticism East and West*, Collier, Middlesex, 1962.
7 Irene Claremont de Castillejo, *Knowing Woman*, Hodder & Stoughton, London, 1973.
8 Irene Claremont de Castillejo, *Knowing Woman*, Hodder & Stoughton, 1973.
9 Richard Parry, *A Guide to Counselling and Basic Psychotherapy*, Churchill Livingstone, Edinburgh, 1975
10 Peter Russell, *The Global Brain*, Tarcher, Los Angeles, 1983.
11 Thomas Mann, *The Magic Mountain*, Penguin, London, 1960.
12 & 13 Irene Claremont de Castillejo, *Knowing Woman*, Hodder & Stoughton, 1973.
14 Thomas Blakeslee, *The Right Brain*, Papermac series, Macmillan, London, 1980.
15 Swami Satyananda Saraswati, *Lectures on Yoga*, First All-India Yoga Convention, Bihar School of Yoga, 1964.

6 *The Conditions of Yoga Therapy*

PEDRO DE VICENTE MONJO

The art of having a long and healthy life is something for which man has been striving since the dawn of civilisation. In the old scriptures, like the Bible for instance, the longest period of human life was estimated to be between 110 and 120 years. In the Vedas, too, we find the expression, 'Let us live a full life of 100 years.'

Even now there are some isolated areas in the world where the lifespan is around 110 years. There are people in Caucasian and South American regions among whom this is common. Studies done on longevity report that, in fact, in hypothetically ideal conditions, anyone can live to a ripe old age, but some attention has to be paid to the environment, diet, physical exercise and so on to make this possible.

In one of these studies, Professor Zaragoza of Seville University has summed up the necessary conditions under the mnemonic DEMCO, which stands for diet, exercise, medical control, control of mind, objective in life. Any person who follows this five-point prescription for a long life could, provided they do not suffer from some unforeseen disease, increase their life expectancy in exactly the same way as those so-called 'special cases' discovered by anthropologists.

On the other hand, every person who has a lifestyle that includes unhealthy habits like drinking alcohol to excess, smoking, over-eating, lack of physical exercise, exposure to constant stress, and who doesn't take any preventive precautions, is certainly going to reduce his life expectancy, or at least his chances of having a healthy life.

But what is health in the first place? According to the World

Health Organisation, health is a positive state of physical, mental and social well-being and not merely the absence of disease or infirmity. This suggests that psychological and mental functions are just as important as physiological aspects when considering the quality of health. And from the physiological point of view, good health could be considered as the capacity for adaptation to all normal situations and activities by people of roughly the same age, culture and environment, with approximately the same lifestyle and habits.

In considering these conditions we will be looking, in effect, at the major physiological systems that are expected to be adaptable and durable during a normal lifetime. Furthermore we will also consider how yoga techniques can improve the general level of health of these systems and hence prolong life.

WHAT IS HEALTH?

First of all, let us define what health is. It could be classified on the following scale:

1 Disease – the inability to perform to total capacity due to the influence of external factors resulting in physical, mental or emotional abnormality.
2 Poor health – difficulty in adapting to change and in following normal activities without incurring physical, emotional or mental distress.
3 Mean health – what is considered normal adaptability within most cultures without showing any sign of stress.
4 Good health – the capacity to perform normal tasks and to adapt, regardless of age, circumstances or environment.

It is not necessary to describe the conditions of disease, for we can quickly summarise the conditions required for positive health as follows and, where these are lacking, then disease, in one form or another, is likely to be present. A healthy person should have the capacity for effort and physical activities; the capacity for mental activity; the capacity for concentration and creativity; and the capacity for enjoying life.

Good health means growth; that is to say, the continuous ability to increase our knowledge and capacity for enjoyment every day of our lives. Poor health means *aging*. It means loss of interest in life, in creativity, activity and so on. However, in order to understand what happens to this diminishing capacity, we have first to become familiar with the role of the cardiovascular and respiratory systems in the process of growth and aging. These systems are affected by what could be called 'risk factors' that accelerate the process of aging, and these risk factors very often cause diseases such as ischaemic heart conditions, myocardial infarction, hypertension, obstructive chronic bronchitis and some spinal abnormalities. According to the International Heart Organisation, these risk factors are:

1 Stress (emotional disturbances)
2 Lack of physical exercise or exercises wrongly performed
3 Smoking tobacco
4 Obesity
5 A high blood cholesterol level
6 Hyperuricaemia (increased uric acid in blood due to wrong diet)
7 Alcohol, drugs, excessive coffee or tea or other stimulants

The way in which some of these risk factors work to induce an unhealthy condition is well known to the public, but there are others, just as dangerous, whose effects are not so well known. So let us take each one and see what it does to the body.

Take the effects of tobacco, for instance. The public in most countries is well educated about its effects as a narcotic, about what it does to the lungs and about its relationship with cancer and heart disease, but how many people realise that it also increases the tendency towards a cerebral stroke, that it badly impairs circulation and can cause Raynaud's disease (progressive numbness of the extremities caused by spasm of the arterioles), or that it causes narrowing of the arteries and high blood pressure?

We think we know about the effects of alcohol, socially and physically – that it can lead to cirrhosis of the liver and alcoholic myocardiopathy – but do we really consider what excessive amounts of tea and coffee do to our health? They can not only alter the heart rhythm and rob the body of vital minerals and vitamins, but can provoke mental disturbance, leading to stress and insomnia.

We have recently become aware of the dangers of over-eating, especially of animal fats. Not only does extra weight mean extra work for the heart but, because of the sheer difficulty of movement with a heavy load, it also puts a strain on the joints, especially the spine. Studies have conclusively shown that people of normal weight are less prone to coronary problems, respiratory diseases, back problems and hypertension.

How many people, too, are aware of the acid–alkaline level of their diet and the effect this has on their health? A high level of uric acid in the blood from eating too much meat or salted or dried fish can cause some crippling and painful diseases like rheumatoid arthritis and heart trouble.

There is no doubt, then, that in order to have a positive state of health, we have to reduce these risk factors and create the right conditions for well-being. Through yoga, as many studies have shown, we can create the right conditions to ensure that we are less vulnerable to disease and that all our physiological functions and systems work normally.

Among the physiological systems that yoga can improve are the cardiovascular, respiratory and musculoskeletal systems. Furthermore, there is a close relationship between all three of these systems, because the working of the heart depends greatly on the oxygen demand from the musculoskeletal system, both while at rest and in action, an adaptation that is not immediately apparent because ventilation is actually carried out by the musculoskeletal system. Thus the function of the heart and lungs depend on the needs or demands of the muscles for oxygen to perform a certain exercise or work. Therefore, the amount of muscle in use, and their control and coordination, will decide more or less the work for the heart and lungs.

Basically here we are speaking about movement, and all movements have important repercussions on the spine. The spine itself is a structure that can increase the work of the heart or lungs, especially if it is not healthy, as, for instance, in the condition known as hyperkyphosis (see Figure 6.1). The deformities in the thoracic cage create tremendous difficulties in respiration, which, in turn, puts strain on the heart.

In fact, there are a number of conditions which alter the normal functioning of the cardiorespiratory system, such as

1 an abnormal state of muscular tone at rest and on exertion;
2 particular types of muscular work or exercise systems;

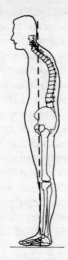

Figure 6.1   Hyperkyphosis, showing deformities in the thoracic cage.

3   certain types of mechanical ventilation;
4   spinal abnormalities or deformities;
5   particular types of muscular tissue and metabolism;
6   coordination of muscular movements.

But perhaps the main culprit in many of these conditions is the inability to use correctly the muscles employed in respiration.

1   The supraclavicular respiratory muscles take in 25 per cent of total ventilation.
2   The thoracic or intercostal respiratory muscles can ventilate 35 per cent of total pulmonary ventilation.
3   The diaphragmatic and abdominal muscles can take in 40 per cent of the total ventilation

Respiratory rate and volume are correlated in such a way that usually the faster we breathe, the more ventilation volume is reduced, as in states of stress, for instance. And the average person does not use the best kind of respiration, i.e. abdominal, while sitting or standing, but only uses the kind of respiration referred to above in groups 1 and 2, which can in turn create postural problems. When standing, for instance, only 35 per cent of the total possibility of effective ventilation is made use of. Yogic training increases the possibilities through correct usage of the clavicular, thoracic and abdominal muscles.

The musculoskeletal system is responsible for the many activities we are able to perform, such as sleeping, resting, sitting, walking, running, dancing, fighting, and so on. In all of these activities there is an intricate relationship between the cardiovascular, the respiratory, the metabolic and the digestive systems, coordinated by the autonomic and central nervous systems. The physiological response is different with every type of activity. For example, when a person is at rest, there is a reduction in heart rate, blood pressure and the need for oxygen. The breathing is also slowed down. In contrast, during physical exercise there is an increase in heart rate, blood pressure and oxygen requirement and consequently breathing is much faster. However, in a healthy person, heart rate and blood pressure are reduced to normal resting rate within a reasonably short time, so avoiding stressful after-effects.

ILL-HEALTH

Ill-health, then, can be seen as a loss of control over these physiological systems. Often, this is caused by loss of awareness. For example, people tend to use more muscles than necessary to perform particular tasks. When we screw up our faces or tense all our muscles just opening a door, when all we really need to do is use our wrist and hand muscles to turn a key, do we realise the adverse effects this has on our nervous systems? Recovering our control through yoga practices has a positive effect on our emotional states. Gaining control over the autonomic nervous system through asanas, pranayamas, bandhas, mudras and kriyas leads to greater sensitivity as to how we use ourselves and our energy, and reduces the need for wasteful muscular or nervous activity, so improving balance, coordination and, ultimately, our general state of health. Muscular control alone leads to the ability to reduce muscular tone and, consequently, anxiety, aggression and metabolic activity are also reduced.

There are several chronic and acute diseases of these three systems that we have been discussing that can be positively helped by yoga practices under proper guidance. Let us now look at some of these conditions and the appropriate yoga methods that can be employed.

Hypertension

There are two types of hypertension, but the essential feature of both is high blood pressure. Unless there is evidence of organic damage, it is usually assumed to be caused by unrelieved stress and is therefore susceptible to certain yogic therapeutic methods. As the arteries are under the control of the autonomic nervous system, but at the same time depend on the activity of the muscles, reduction of muscular tone and emotional control can be obtained with pranayamas and some asanas. Since the aim of the treatment is to obtain arterial dilatation and reduction of arterial spasm, savasana is particularly effective – 10 minutes of practice daily can reduce blood pressure by 20 per cent. (It should be noted here that this does not mean that people with normal blood pressure will be adversely affected by this pose; on the contrary, it helps to maintain homoeostasis throughout all the systems of the body.)

One of the main dangers of hypertension is that it can cause myocardial infarction and brain strokes, so some asanas like ardha halasana – especially if it is done progressively, starting by raising one leg at a time – gradually retrain the arteries to take increased pressure and blood flow. This can also be aided by the use of a tilted table, like the one I saw used at Kaivalyadhama Yogic Hospital in Bombay. By tilting the table at progressively increased angles so that the legs are eventually higher than the head, the patient's blood pressure can be controlled. This has also been tried with patients recuperating after a stroke, but it should be made clear that this practice carries some degree of risk for these patients and therefore the technique should *always* be carried out under medical supervision.

Venous and lymphatic insufficiency

These are common conditions that are also responsive to inverted poses. Inversion helps to drain blood that has pooled in the lower parts of the body. In certain asanas there is a kind of muscular massage on the venous walls that helps to release blood so that, with practice, a person suffering from this condition can even sit cross-legged without any ill-effects.

Peripheral artery disease

This is another common condition in which arterial blood flow is reduced in one area, usually the limbs or the head. Here again inverted poses are effective. In the upside-down position, gravity increases blood flow through the brain arteries, helping them to become stronger and more elastic. It is also important to have a balanced system of movements and poses so that the body is adequately exercised, so improving circulation as a whole in a remarkably short time. However, certain precautions *must* be observed:

1　Do not practise inverted poses when there is high blood pressure, headache or conjunctival congestion.
2　If there is an uncomfortable sensation of heat or pressure building up in the head, stop the posture immediately.
3　Never try to get up immediately at the end of the asanas or you may experience dizziness.

Chronic bronchitis and asthma

These are two common respiratory conditions in which emotional states alter breathing capacity. While emotional control can be helped by savasana, pranayama techniques have a direct effect on respiration and therefore have the following advantages:

1　Pranayama is controlled respiration with self-awareness, so increasing sensitivity to alterations in breathing rhythm, especially while under stress.
2　Pranayama increases the volume of oxygen in the lungs while slowing down the rate at which it is inhaled.
3　Pranayama prolongs exhalation and induces calmness and self-confidence.

It is also useful to practise some of the kriyas to cleanse the upper respiratory tract, which also improves breathing.

Emphysema

This is a condition in which lung tissue has been destroyed by the inhalation of certain kinds of dust or allergic substances, and it is extremely hard to treat. However, like all respiratory conditions it is important to reduce anxiety through relaxation as well as controlling unnecessary muscular activity. It is also important to have a clear picture of the person as an individual and to understand how his or her limitations, such as stiffness of the chest or spine, may be affecting the condition. Postural deviations like kyphosis or lordosis can seriously impair a person's ability to breathe adequately but this can be helped by careful instruction in progressive movements and adapted asanas.

Sinusitis

This is usually caused by inflammation of the nasal sinuses from a bacterial infection. The pus that accumulates in the cavities can both prevent proper breathing and cause severe headaches. A most effective method of clearing the cavities is to wash them out by means of neti kriya to prevent the condition developing in the bronchial region, as well as preventing it from developing into an allergy.

Heart failure

This is the name given to the acute or chronic condition that occurs when the heart is unable to supply sufficient oxygenated blood to the tissues, resulting in the accumulation of fluid in various parts of the body, depending on which side of the heart has failed to maintain adequate output. It is usually treated with diuretics to release the fluid and with digitalis to stimulate the action of the heart. Shortness of breath is a distinctive feature if the condition is well established. The complaint can be successfully treated with the application of some pranayama exercises such as ujjayi. Reduction of anxiety is, of course, of paramount importance, and therefore savasana and other asanas should be utilised if the patient is able to perform some simplified versions. Abdominal breathing performed while sitting (sometimes lying down in relaxation is impossible at first because of breathing difficulties) helps to improve ventilation and the blood flow through the pulmonary vessels, as well as inducing calmness and control over respiration.

Ischaemic heart disease

This occurs when narrowing of the arteries of the heart has put severe stress on heart action, especially during any physical exertion. Patients suffer from severe chest pains, not only because their lifestyles are inappropriate but also because the illness itself causes further distress through the impending threat of a heart attack. So it is important for the patient to learn:

1 emotional control to avoid coronary spasm – savasana, pranayama, kriyas and selected asanas such as simhasana are helpful in improving emotional control and stabilising respiration;
2 muscular control to reduce oxygen needs while at rest as well as during effort through selected practices.

Summary

These are by no means all the conditions that can be treated quite successfully through yoga therapy, but perhaps they are the ones best known to the general public. The important point to remember about all of these conditions is that all practices must be performed with great care so that the patient feels confident that the techniques will help his or her condition and is not worried about any possible ill-effects, since this would only add to their existing anxieties. It is therefore necessary for the therapist to see that the patient performs the yoga practices in a way that suits individual capacity and limitations. It is then easy for the patient to see how progressive application brings greater well-being and freedom from pain.

Not only does yoga release people from the suffering caused by the chronic illnesses discussed here, but it also enables them to adapt their lifestyles in a more realistic way once they have recovered sufficiently, so creating the possibility of a long and healthy life.

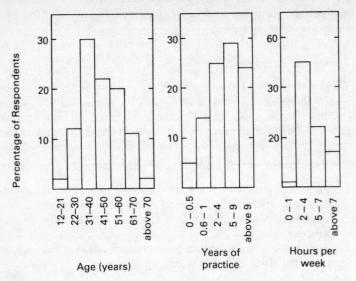

Figure 6.2 Analysis of 2,672 respondents to survey on results of practising yoga.

USE OF YOGA – TWO SURVEYS

The first survey is based on data from 2,672 questionnaires completed for the Yoga Biomedical Trust. Preliminary analysis of these results was published in *Yoga Today* (Vol. 9, no. 5/6, Sept./Oct., 1984).

Figure 6.2 gives an age analysis of the respondents to the questionnaire, and indicates their experience with yoga and the hours devoted to yoga each week. Table 6.1 shows that the majority of people felt that yoga had improved their sense of well-being, energy level and capacity for work. A large proportion felt that yoga reduced their days off work and consultations with the doctor. At least half observed that their resistance to common ailments such as colds had improved and their dependence on general medicines and tranquillisers and sedatives had been reduced. Table 6.2 gives a breakdown of specific complaints that yoga had helped alleviate.

The second survey is based on 4,253 cases treated in one year at

Table 6.1   Effect of yoga on respondents

	Increased %	Decreased %	No change %	N/A %
Energy level	69	1	24	6
Working capacity	55	2	33	10
Susceptibility to colds and 'flu	3	45	31	21
Days off work on sick leave	1	23	35	42
Consultations with doctors	2	32	40	26
Consumption of medicines	1	30	30	39
Consumption of tranquillisers or sedatives	1	22	20	57
General sense of well-being	89	1	6	4

Table 6.2   Only those respondents are included who answered 'Yes' or 'No' to the question 'Has yoga helped?'

	Cases	% Helped
Back disorders	1142	96
Asthma or bronchitis	228	88
High blood pressure	150	84
Heart disorder	50	94
Duodenal ulcers	40	90
Haemorrhoids	391	68
Diseases of nervous or muscular system	112	96
Cancer	29	90
Diabetes	7	86
Rheumatism or arthritis	589	90
Pre-menstrual tension	848	77
Other menstrual disorders	317	68
Menopause disorders	247	83
Obesity	240	74
Migraine	464	80
Insomnia	542	82
Excessive anxiety	838	94
Heavy smoking	219	74
Alcoholism	26	100

Table 6.3 Diseases successfully treated by yoga

Disease	Average number of patients treated		Average number of cases per year			
			Relieved or improved		Not relieved	
	No.	%	No.	%	No.	%
Respiratory disorders	156	25.5	142	91	14	9
Alimentary disorders	146	23.9	133	91	13	9
Circulatory disorders	33	5.4	26	78.8	7	21.2
Metabolic disorders	99	16.2	83	83.8	16	16.2
Rheumatic disorders	20	3.3	17	85	3	15
Genito-urinary disorders	23	3.86	15	65.2	8	34.8
Psychic and nervous disorders	104	17.0	82	78.8	22	21.2
Miscellaneous	30	4.9	27	90	3	10

the Kaivalyadhama ICY Health Centre, Bombay. Table 6.3 lists those diseases, common to all big cities, that were successfully treated by yoga. The table demonstrates that the largest number of patients requested help for respiratory disorders caused by industrial irritants – dust, fumes, smog and other pollutants – followed by alimentary/digestive disorders, probably caused by stress, and psychic and nervous disorders.

7 *New Frontiers in Yoga Therapy*

BARBARA BROSNAN

Increasingly over the years, the practice of yoga in various aspects – asana, breathing, chanting, relaxation, meditation – has been tried out for and by people with a handicap. To start with it was little more than someone with arthritis, or possibly somebody who had had a stroke, being integrated into an ordinary yoga class. They did what they could, remained quiet while others were performing postures they were unable to attempt, but joined in, naturally, with pranayama, relaxation and meditation.

The picture now is vastly different. There are classes solely for people with physical handicap or mental handicap, for the over-60s, for the frail elderly, for the over-80s, for the blind, for the deaf, for the almost – or quite – helpless.

The development has not been rapid. Teachers cautiously took on one or two people with more pronounced disabilities, perhaps on an individual basis, then started classes for 'multiple sclerosis sufferers', for 'people with cerebral palsy', 'those with spina bifida', and so on. They were both amazed and stimulated by what happened, by the progress people made, by the tremendous value even those with the heaviest handicap found they received from the classes – as did the teachers themselves.

This overall value of the practice of yoga for handicapped people – no matter how great the degree and/or type of disability – is most striking, for where there is handicap the integration within the individual of mind, body and spirit is often very incomplete. This applies whether the disability is acquired or congenital. It is here that yoga, with its emphasis on concentration, with its practice of 'thinking through' a posture as if it were being perfectly accomplished, can help. As the student thinks of movement into and along a paralysed limb, of flexion into fixed joints, of movement in a rigid spine or around a locked shoulder or hip girdle, miracles do not occur; actual movement may not occur at all, or at best minimally. But the mind and body integrate within their common purpose, as witness the common remark of a handicapped person: 'I feel so much more of a real person. I feel all one now.' What better description could there be of yogic integration?

The use of the breath to facilitate movement often does almost produce the 'miracle' and results can be attained that were never dreamed of. For example, when a 'non-walking' person with cerebral palsy and gross athetoid movements walks, it feels miraculous, despite the fact that it may have occurred after some years of assiduous yoga practice. Yoga is *the* therapy for people with handicap, because of its involvement of the *whole* person, because it is for anybody, whatever their age or type of disability, and because it requires no expensive outlay or special apparatus and not even much space.

While integration, the one-ness of body, mind and spirit, is the aim for all handicap, there is a slight difference of subsidiary aim when considering those with acquired handicap – perhaps one that is deteriorating – compared with a congenital handicap. Where the handicap is acquired, the aim is to use yoga to restore an ability that was previously there and/or to forestall deterioration. With congenital handicap, yoga serves to produce, or control, a movement not previously there, or that is wildly uncontrolled, as well as forestalling the very rapid deterioration that may occur in later years. Here the yoga therapist may not only have to start from nothing, that is with a totally helpless student, but from a positive disadvantage; the student may have constant athetoid movement, so there is total lack of control.

A yoga therapist is not, or should not consider him/herself, faced with a person with a disease but by a person with a symptom, or at least with something out of the ordinary that affects movement. The student may be deaf, blind, dumb, but this presents no more than a simple teaching problem. Such a person can be accommodated in an ordinary class, unless two or more of these conditions are present in one person.

Certain types of handicap have certain presenting symptoms with which the therapist becomes familiar. For example, weakness of movement in one, two or all limbs, or even paralysis, is the presenting symptom in diseases of the central nervous system. Whether the person has multiple sclerosis, Parkinson's disease, cerebral haemorrhage, post-polio encephalitis, motor neurone disease, cerebral tumour, muscular dystrophy, spina bifida, Friedreich's ataxia, cerebral palsy or a broken neck, these all represent movement limitation and the therapist will work accordingly. Postures will be modified, particularly the starting point, for this position may well be an impossibility. For instance, when people cannot stand, trikonasana can be performed lying on the floor. Sometimes the therapist will then restrict movement where it is easy and demand more work where the movement is difficult; where there is rigidity or tension, far more attention should be paid to relaxation and breathing. In fact, months may be spent 'mastering' these two aspects before even 'thinking' into postures.

Diseases of the skeletal system present limitation of movement and poor posture, which may have to be corrected over and over again, standing, sitting, lying. The therapist working with people with arthritis, flat feet, knock knees, congenital hip dislocations (perhaps treated in childhood but leaving residual problems), sciatica, disc lesions, ankylosing spondylitis, has to isolate certain groups of muscles, restricting some to allow others to work; has to work on stretching to open up disc spaces and relieve the nipped nerve; and has to teach relaxation to remove muscle spasm which is restricting joint movement.

Disorders of the respiratory or cardiovascular systems will present problems of breathing. If severe, the problem will be present even when the person is at rest, whereas other individuals may only experience breathing difficulties in the asanas – changes in the breathing pattern, distress, gasping, ragged breathing, are all signs that the demand on the heart and lungs is excessive. The presenting symptoms may range from congenital heart disease, congestive heart failure and rheumatic heart disease to anaemia, abberrations of blood pressure, bronchitis, emphysema, asthma or cancer of the lung. But whether the disease is respiratory or cardiac in origin, the therapy required is steady work on elementary breathing exercises – not elaborate pranayamic techniques – in order to improve lung potential, and simple mild asanas to strengthen cardiac muscle and cardiac output. Progression is dictated by response to the techniques within the limitations imposed by breathing ability. As breathing improves, breathing exercises give way to pranayama; as heart functioning improves (the two will most likely change *pari passu*) so stronger asanas are used and held for longer.

When there are disorders of the digestive system, then pain, constipation, diarrhoea and flatulence are the presenting symptoms. The therapist's task here is to work on relaxation and on stimulation of the gut by suitable asanas.

Elderly people may be frail, stiff or simply unfit. The therapist should handle these people in exactly the same way as those with cardiac or respiratory symptoms, but expectations should be realistic. Though the results of yoga can be 'magical', an 80 year old cannot be expected to have the stamina of a 20 year old. However, a good deal of stamina, considerable flexibility and a wonderful sense of well-being can result from work with the elderly.

The yoga therapist may be presented with quite a wide range of symptoms in handicapped students. These can include obesity, malnutrition, eczema, anxiety, tension, epilepsy and all forms of cancer. It is essential for the therapist to remember that yoga can help because it is wholistic – the person is treated, not the symptom or disease. Yoga integrates – and the integrated person can *cope*.

It is not possible to work with disabled students in the same way as with the fit. Work must be done with small numbers, often purely on a one-to-one basis and sometimes even on a two-to-one basis; that is, a teacher plus one helper for each student. In this case, the helper immobilises or holds the person while the teacher works in the positive direction. Much work can be done in a chair, or even in bed, but the floor is infinitely preferable because of the freedom and stability it provides.

Postures are not considered in their complete form but reduced to absolute basics and worked on piecemeal. Let's take paschi-mottanasana, for example. The therapist may concentrate on helping the student to straighten the knees *or* on getting the correct movement from the base of the spine *or* on getting the arms fully extended above the head. Each one of these aspects may have to be worked on for some time, totally ignoring the others. Start on what you can get and work on it is the axiom.

Where the emphasis in postural practice is wrong and creating imbalance, the strong area must be 'taken away'; that is, made unable to function so that the region being worked on can be strengthened. This can happen in people with cerebral palsy. They may have a tendency to use strong, jerking movements of the head, neck and shoulders instead of the back to achieve spinal flexion. For example, a young spastic unable to press up on her hands in bhujangasana because of flailing leg movements was made to sit on her haunches. By 'taking away' the legs, she was then able to work with her arms and strengthen her back muscles through this adapted posture.

The right attitude of both therapist and student is essential to the whole process. The student *must* show willingness right from the outset. Commitment and belief are also necessary for good results, but emerge in their own time. Enthusiasm can be helpful but if there is too much excitement or euphoria it can swing in the opposite direction; the therapist must therefore be careful not to build up enormous hopes or promise miracles. It is also important not to overload the students with too much philosophy. Where there is physical disability, physical work is needed, while the mentally handicapped want fun, stimulation, achievement and success. However the totally helpless may sometimes respond more to the spiritual aspect of yoga, and it is not unusual for people to be attracted to yoga in the first place by the spiritual possibilities, whether they are fit or unfit.

The therapist should exude a spirit of calm optimism, a conviction that progress *will* come, that yogic integration *will* take place. Unhurried quiet concentration from both teacher and helper creates the right atmosphere for work. At the same time, humour is a vital ingredient: no good comes from everything being deadly serious. Another vital quality is the capacity for acute observation. The therapist needs to be able to spot the flicker of movement, the merest suggestion, the tiny clue that points the way to possible adaptations for the rearrangement of a posture. For example, noticing a young student with spina bifida handling his own lower limbs to move from bhujangasana to sitting up once produced the idea that it should be possible to perform parvatasana balancing on both legs together as on a pivot.

Ideally it is best if the student can be where yoga is a way of life. The student organised into a daily routine geared to yoga, surrounded by people on the same 'wavelength', with similar interests and eating a yogic diet can benefit enormously. Unfortunately, this is rarely possible. However, that the therapist should be wholly yoga oriented, passionately convinced of its value as a therapeutic tool – as *the* therapeutic tool – is a *sine qua non*.

Immense patience is also necessary – the ability to spend session after session working for the minutest bit of improvement. Ingenuity is also needed to devise ways and means of reaching the goal. Intuition – the combination of a seeing eye, a lateral thinking brain and an ability to focus entirely on one aspect of the problem, symptom or deficiency – is of incomparable value. To this extent one might claim that the first-class yoga therapist is born, not made. And it is true that some therapists have greater natural affinity for the work than others, but all need special training.

To start with, a good working knowledge of anatomy and physiology and of the basis of each disability is required. Here it should be stated that medical knowledge, though helpful, is not actually necessary, only awareness of what results physically from the medical condition.

A knowledge of psychology is also useful. An awareness of personality types – of the type to push and the type to restrain – of complexes, of guilt factors, of all the psychological factors that can hinder progress, is vital. The ability to detect tensions and strains, which the individuals themselves may be totally unaware of, enables the therapist to direct the work with love and compassion and without useless clogging sympathy. And of course, they also need to be able to stand back and make students do things for themselves. They need to have the ability to encourage warmly, to know when to drive and exhort and when to let people lie fallow and grow within before progress shows outside.

The best training, of course, comes from the students themselves. A receptive therapist learns from the people he/she is working with, by observing them day after day. And it *is* the situation itself that teaches, for even though the basics – anatomy, physiology, elementary psychology, the art of teaching – have been learned in formal training, there is still a lot more to know about what it means to live with a disability. The therapist has to put him/herself in the disabled person's place and learn how to help in constructive ways – learning to turn a helpless person over, learning how to move them from lying to sitting, where to put the hands, where to support with the leg. The therapist not only has to learn how to break down postures into simple steps, but how to move a student, for example, from the prone position to shashankasana, or what movement to use to bring them into marjariasana; how to take a person from shashankasana to parvatasana or up into sarvangasana. These techniques can be learned through specific words and *must* be learned for the sake of the handicapped person.

A good therapist will also liaise with other therapists, perhaps with a physiotherapist for help in movement or with an Alexander teacher for guidance on posture, strain and tension, but he/she should not lose awareness that yoga is beyond physiotherapy or whatever bodily technique is used, because it is therapy of the body, mind and spirit.

One of the problems with disability is the fact that the person has the feeling of there being no hope. The medical profession tends to diagnose, label and leave it at that. They may be supportive, providing physiotherapy or welfare services, but the individual (and the family) have to 'learn to live with it'. The spirit of optimism is missing, especially where congenital or progressive degenerative diseases are involved. And personal outlook is the most important factor in determining how someone with a handicap is going to progress in life. The medical profession talks glibly about 'coming to terms with disability', of 'accepting things as they are', but the situation is much more complicated than that.

Disabled people are human beings who respond in individual ways. There are some who are determined to be like everybody else, who resolve to do absolutely everything or 'die trying', but by the time all their needs are met, they may find their lives (or what little is left over) not worth living. Some go to the opposite extreme; they give up, go 'vegetable', or wear a permanent chip on their shoulder, wasting endless time moaning 'Why me?'

The therapist's approach, and the student's response, must be three-pronged – physical, mental and spiritual. The emphasis will vary from person to person. With some the emphasis will be on the physical side, involving the breath. When positive results have been established, emphasis can then be transferred to the mental, using the power of the mind through concentration to facilitate physical work. From this the next step will be to meditation and the opening up of the whole spiritual field. However, it will sometimes be necessary to start with the mental approach – the power of the mind over the body, then the stilling of the mind, the use of relaxation and then on to the physical; with the mind and body in harmony, it is finally possible to expand into the spiritual aspects. With a few people, the approach comes through the spiritual approach – the experience of meditation, its effect on mind and body and from this to the use of the breath and then, lastly, involvement of the body. But whatever the line of approach, the result will be increased wholeness of the person and a dynamic, positive approach to life which will improve immensely the quality of the life lived, *regardless* of the degree of disability.

Yoga therapy for the disabled is a new frontier that requires further exploration, but the results of small groups throughout the country show that yoga as part of a rehabilitation programme can yield positive results. If yoga were to be integrated into hospitals and clinics, the disabled would have greater hope for the future. Learning of its potential (preferably with personal experience) should be part of any medical school curriculum for doctors, nurses, physiotherapists and even occupational therapists. Scholarships and facilities for research into the full physiological implications of relaxation, meditation and breathing techniques would result in a diminution of crippling degenerative diseases and a better quality of life for the handicapped. All members of the caring professions should be orientated towards viewing yoga as a valuable therapy, encouraging it and supporting its practice.

8 Relaxation and Meditation: Borrowing and Returning

MARK BLOWS

In city life in most countries there is an ever-increasing amount of stimulation and pressure towards activity. This leads to excess mental arousal, cycles of unproductive thinking, and over-excitation of the endocrine system. When unrelieved, stress aggravates anxiety and susceptibility to disease.

Sometimes the stress is readily apparent to us because of subjective feelings of distress; sometimes it does not become apparent until serious symptoms occur. Many city people even become adapted to high levels of stimulation, are bored without it and may compulsively seek it without being aware of its harmful effects, such as loss of subtlety of perception and various stress symptoms. Stress is reduced when we have opportunity to control and regulate our activity according to our individual rhythms of exertion and rest. Recent research links the harmful effects of stress with loss of control over pressures, external coercion and uncertainty.[1]

There are two general practical approaches to reducing excess stress: environmental change, such as better working conditions, less crowded public transport and so on; and changes in the attitude and resources of the individual. The two approaches are not altogether separate, for a more resourceful, calmer person is likely to be more open to constructive environmental changes. According to Fritjof Capra, the best basis for self-help preventative health care lies in a combination of practical stress-reduction techniques and an overall balanced pattern of living.[2]

I have selected some such practical techniques for discussion. Three of them will be discussed in detail: Transcendental Meditation (TM); Clinically Standardised Meditation (CSM); and a third method developed by a colleague and myself, which for convenience I have called the 'Farmer and Blows Method'. Preventative health care or self-development and instrumental applications in therapy will be considered. The Farmer and Blows method has advantages over the other two methods for therapeutic applications, and its historical source is different from them. In their general preventative use, however, all these methods provide rest and time for restorative functions to act within the nervous system — something which is difficult to obtain in a society alienated from the rhythms of nature.

In order to avoid confusion, two distinctions are necessary — between meditation in modern psychological practice and in religious practice on the one hand, and between psychological practice and popular goal-oriented practices on the other. The first distinction is best understood in an historical context. In drawing this distinction, the two techniques, TM and CSM, will be described. After that the Farmer and Blows method will be described in its historical context separately. Examples of practical clinical applications, using various methods, are provided towards the end of the chapter.

TM AND CSM

The term *meditation* gained currency in Western countries with the spread of TM, introduced by Maharishi Mahesh Mahayogi, head of the International Meditation Society. Patricia Carrington, psychologist at Princeton University, USA, continues to use the term in her modification of the TM technique, 'Clinically Standardised Meditation'.[3]

Carrington calls the TM type of procedure a centring technique. Originally a general orientating and preliminary step, it has been abstracted from traditional meditative practice. It would be surprising if this type of technique were sufficient for attaining full samadhi or higher states of consciousness as experienced by yogis. It may, on the other hand, be used as a first step towards further development, as, for example, the TM system of siddhi training. Both TM and CSM involve the soft repetition in thought of a mantra, usually a Sanskrit word or phrase derived from traditional Indian sources.

In India, to utter a mantra means to partake of the characteristics of a God, for the mantra is associated with the name of a deity. To the Westerner using TM or CSM the mantra is a pleasant, unique, personal sound, but vague in meaning. The mantra is kept a secret, chosen for the initiate by his teacher in the case of TM and selected by the client from a list in the case of CSM. There may be sound reasons for preserving the uniqueness of the mantra as a secret, such as enhancing dedication and preventing people teaching others inappropriately. The International Meditation Society have not, as far as I know, explained the need for secrecy, but secrecy is, of course, quite usual in any ancient *rite de passage*. It is fully normal in that context and may attract people who feel a lack of mystery in everyday life.

Carrington is sceptical but still encourages the client to preserve secrecy of the mantra for their own use so that it is a 'special word for turning inward' (page 172 of *Freedom in Meditation*). She has added a small number of words that are not of Sanskrit origin and that list has been published.

There are several difficulties with TM. Although the International Meditation Society assert that the practice is not a religious one, several incongruities remain. There is, in my opinion, ambiguity and distinct cognitive dissonance in the way in which the TM method is presented. According to Byron Rigby, a psychiatrist and 'chief minister' in the Maharishi's system of 'government':

> Transcendental Meditation is a very simple, natural, mental technique which is practised for twenty minutes morning and evening, sitting comfortably in a chair, and its effect is to expand consciousness and release stress from the nervous system. It does not involve any kind of philosophical belief or membership of any group, or any sort of religious commitment. It is a completely mechanical technique which has the effect of producing a settled state of consciousness in the physiology.[4]

After making such statements about their technique, the TM Organisation act to recruit members for their system, aiming to establish an 'enlightened' international community who are in favour of world peace and other desirable objectives. To tell initiates that TM does not involve 'any kind of philosophical belief or membership of any group or any sort of religious commitment' is controvertible. The glibness of the description 'mechanical technique' invites challenge. Are Westerners so hard-boiled that they will only buy a 'mechanical technique'? Many are attending Hindu and Buddhist centres where there is no ambiguity about, or apology for, the orientation towards religion. And indeed, my TM friends tell me that TM has its origins in the teaching of Sankaracharya, a Hindu philosopher, after all.

Carrington, apparently dissatisfied with the psychological technique of progressive relaxation as used in Western clinical practice, developed CSM, which is similar to TM but more suitable for use in medical and psychological treatment and research. She puts forward two objections against the use of TM in a scientific context. Firstly, the induction and the details of practice are secret. Secondly, no research findings using TM may be published without the permission of the International Meditation Society. Thus a basic ethical principle of science is violated, under which knowledge is public and available for testing by anyone qualified to do so. Carrington has thus carried the secularisation of 'meditation' as a psychological practice a stage further, clarifying the distinction between meditation as used in religion and as used in psychological practice. (Benson preceded her in arriving at a secular form of meditation, developing another variant of TM. He substituted the word 'one' for a mantra and made other changes.[5] Nevertheless, Carrington's method is closer to the original.)

In popular usage the term meditation is used to refer to methods of relaxation, although it may also be confused with prayer, hypnosis, psychocybernetics, autosuggestion and similar activities which tend to be goal-directed; either covert or frank strivings towards a goal tend to be confused with elements of relaxation. The three methods discussed here, TM, CSM and the Farmer and Blows method of relaxation, all involve a process of 'letting go', detachment from strivings and a temporary dissolution of structured thought. Deep rest becomes possible, allowing restorative activity to proceed within the central nervous system. After meditation, practical problems which have been preoccupying the mind may be seen in a fresh light, unencumbered by automatic cycles of thought.

To illustrate the converse of this concept of 'letting go', one of my clients told me that her attempts to meditate had become fruitless. Estranged from her husband, she let slip into her meditation a prayer, expressed in a poetic form but implying that her husband's affection for her would be restored. This perpetuated her main striving, and destroyed the trend to 'let go' and bring her mind to a state of rest.

It is difficult, at least at first, for some people to accept the idea that there is something to be gained by not trying. In many popular forms of psychotherapy, the client projects himself into an imagined situation, achieving a goal. This may add to people's confusion. The methods described here help to check the experience of pressure towards activity in a mind that is over-stimulated.

After the initial training, the three methods, TM, CSM, and Farmer and Blows, are self-administering, enhancing feelings of independence and wholeness and permitting the resolution of stress-related problems without the use of invasive methods, such as the administration of medication (although this does not mean that medication can be avoided in every case).

Distinctive physiological and neurophysiological correlates or indexes have been established for the meditative state, as distinct from sleep and ordinary rest. Parasympathetic activity is increased as against sympathetic activity. One of the most impressive indexes is synchronised alpha rhythm on an electro-encephalogram. For a fuller description of these indexes, I recommend Pelletier's books,[6,7] such as *Mind as Healer, Mind as Slayer*, and for a critical summary of the research issues, I recommend chapters 3 and 4 of Carrington's book.[8] The Farmer and Blows method has not been submitted to physiological investigation to see whether it gives rise to the same indexes, but I trust that it soon will be. My impression from clients' reports is that it would give rise to identical indexes in many cases.

THE FARMER AND BLOWS METHOD

The method of induction for the Farmer and Blows method is different from that of TM and CSM. It is a reliable method for teaching which lends itself well to instrumental applications. Like the other two methods, it has origins in yoga and it has features in common with savasana. This exercise or asana is generally practised in Western yoga classes with a form of progressive relaxation (see *Yoga Over Forty*,[9] Chapter 10, Asanas; Hittleman's book,[10] Chapter 17, Deep Relaxation). As described by Dr Bhole[11] of Kaivalyadhama, it entails lying in a supine position, an awareness of breathing movements, an awareness of the touch of air within the nostrils and eventually, with practice, a passive observation of thoughts and a letting go of the 'internally triggered' cycles of mental activity.

In general, deep relaxation is not easy to teach, but it is easier with progressive relaxation. This in turn is facilitated when muscles are tightened then relaxed. A derivative exercise, involving the alternate tightening and relaxing of a series of muscles throughout the body, has been known as progressive relaxation, a time-honoured technique in English-speaking countries and still used in the treatment of mental disorders. As originally introduced into Western therapeutic practice from South Africa by Jacobson[12] it was a very lengthy, cumbersome procedure, involving the tensing and relaxing of each group of muscles in the body in turn and requiring a total of almost 60 1-hour training sessions. (The connection between savasana and Jacobson's progressive relaxation is rarely made, although it seems obvious that the large Indian minority had cultural influences on the South African population as a whole. I suspect that the political situation and the cultural-intellectual assumptions of the 1930s led to a suppression of this connection.)

Wolpe, also in South Africa, shortened the procedure to four to six sessions, still involving alternate tensing and relaxing of groups of muscles.[13] However, dissatisfied with this, Dr Ron Farmer developed a still shorter method, introducing other principles consistent with both yoga practice and learning theory from modern psychology.[14] I have extended this method in the direction of meditation, re-introducing the practice of passive observation of thoughts or letting-go cycles of thinking which is a feature of savasana, a fact unknown to either of us at the time of developing the method during the 1970s. My extensions were influenced by TM.

It is apparent in the Jacobson and Wolpe methods that the state of relaxation is continually disrupted by the actions of tensing. This may not matter in certain specific applications such as psychoprophylactic training for reduced pain in childbirth. In that example, the pain of contraction disrupts the relaxation state and it is necessary to accommodate to that. In other applications of the method, however, continued tensing is an impediment. It is not surprising, therefore, that progressive relaxation does not always yield the same neurophysiological indexes as meditation.

Farmer solved the problem of disruption using the principle of stimulus generalisation that he derived from learning theory. This solution occurred to him one day when he was recuperating after an appendix operation with time to reflect – a seemingly very simple solution after the event of discovery. Instead of tensing a series of muscles in the whole body, the client does preliminary training with the dominant hand, tensing only that part of the body and allowing the experience of 'letting go' to generalise to other parts of the body in turn, with no more actions of tensing. Once commenced, the relaxation state is enhanced, not disrupted any more.

The exercise is begun by saying the word 'relax' softly on each relaxation. This is introduced in two simple steps so that, without being actually told to do so, the client concentrates on his breathing and prepares for the word 'relax' to become an affective signal within the brain, allowing him to relax quickly in everyday situations. Tensing of the dominant hand is introduced coarsely at first in one cycle of breathing, then extended to gradual tensing and gradual releasing over a number of breaths, becoming progressively finer. In the last stages of this preliminary procedure the client is asked to imagine that, after his hand feels fully relaxed, a little more tension flows away each time he breathes out, all the while thinking 'relax'. Saying 'relax' aloud at the beginning of each breath enables the psychologist to monitor coordination with breathing and it also helps set the habit. By gradually refining the contraction and release of the hand, commencing with coarse tensing that can be felt easily, the way is prepared for generalised release of tension in other areas of the body. Imagining further release of tension enhances this capacity. The instructions include the statement that effort is not required, only letting go. Each section of the body is then covered, using commonsense rather than strictly anatomical sections, consistent with people's ordinary body images. Instructions continue to suggest that the client let go any tension that might be there, and the whole process is monitored through the psychologist's observation and the client's self-reports.

Further instruction is available for people who have problems with particular body areas that resist relaxation. Farmer also took the precaution of including an additional section of pleasant imagery of the client's own choice, in order to cater for those whose anxiety signs or symptoms are vascular rather than muscular. A suggestion of heaviness of the limbs, as in Schultz's autogenic training, was also added. (Schultz's autogenic training, developed in 1926, serves a similar function in Germany and other European countries to progressive relaxation in England, North America, Australia and South Africa. It is also related to yoga techniques and it has become common practice to combine it with biofeedback training.[15])

In my experience, two training sessions are required before the client can reproduce the exercise reasonably well at home. I always prepare the client well for dealing with common difficulties when they are at home. I recommend doing the exercise either once or twice a day, at a regular time other than just after a meal, for general purposes, and at least three times a week for temporary use for specific instrumental applications.

With modifications, the original Farmer method can be extended to arrive at a state similar to that attained during TM or CSM. People may let the word 'relax' become softer and softer until it becomes an effortless beat, at the back of the mind as it were. I suggest, after several sessions, that they stay with the state of relaxation after completing the muscular release. When thoughts arise, as they commonly do, they are not to be fought or suppressed but allowed to float by. Fantasies or images are to be regarded similarly after dominant hemisphere (non-verbal) imagery emerges. If the person has become distracted by the thought or imagery, perhaps having given way to the feedback process of directed thinking, then they gently bring themselves back to thinking 'relax' when they become aware of the distraction.

Some clients learn very quickly. Many arrive at deepening states of relaxation with practice and give subjective reports similar to those given by people using TM. Recently, having taught people the Farmer and Blows method, I have taught CSM as well to some so that they may genuinely choose what suits them. I have found that the majority prefer to stay with the Farmer and Blows method. While the numbers so far are insufficient for a genuine comparison, this lack of preference for a

TM-type method is quite a contrast to results referred to by Carrington, where people preferred TM to progressive relaxation.[16]

It would seem, on first acquaintance with the Farmer and Blows method, that it is suitable for people who have muscular signs of anxiety, whereas CSM is suitable for people who have racing thoughts. This distinction, however, is unwarranted. I have never regarded the Farmer and Blows method as limited to muscular relaxation and have always explained to clients that we are aiming for calmness of mind through a bodily exercise. Dr Jagdish Dua, a colleague at the Department of Psychology, University of Armidale, has commenced comparative studies using meditation and relaxing techniques.[17]

Since the Farmer and Blows method has a range of components, its use is unlikely to be restricted to particular types of disorders. With proper monitoring, it is a safe method to teach, allowing flexibility according to the client's response. I have had very few difficulties with it. Occasionally a client will revivify a past trauma, and judgment is necessary as to whether or not this is a helpful experience. In some cases it is better to leave the exercise at the earlier muscle relaxation stage or change to a form of pranayama, along the lines recommended by Dr Gharote.[18]

It is a popular practice to purchase sound cassettes for self-instruction. However the use of cassettes such as those issued by Carrington for CSM and Farmer for a variety of purposes has some disadvantages, for, having learned with a cassette, people tend to rely on this external device, failing to trust their own internal cues. Furthermore, cassettes give a false impression that skills of gaining control in the central nervous system may be acquired instantly. As with all skills that are worth having, patience and discipline are essential. I would therefore only recommend tapes in cases of geographical isolation or other exceptional circumstances.

Farmer has developed a whole series of meditative-like procedures for specific purposes, and I see no objection to a person extending their skills through listening to tapes, if they have learned a core skill through relying on their own internal cues, but it must be remembered that there are catalyst elements in personal communication with a teacher that cannot be replaced by tapes or by books.

The Farmer and Blows method has proved useful in a surprisingly diverse range of instrumental applications, including specific phobias, social anxiety, essential hypertension, sexual inhibition, tension headaches, migraine, insomnia, menstrual pain and various specific stress symptoms. It increases the effectiveness of systematic desensitisation where feared situations are broken down into graduated steps and, with training, the word 'relax' enables a person to lessen sympathetic nervous system arousal quickly. That is best done on recognition of minor signs within the body, forestalling the accumulation of tension. The method is not usually used alone in the instrumental applications, but in combination with other procedures; it may be regarded as a useful basic skill, like woodwork joints in carpentry. There is much more to houses than woodwork joints, however. In psychology and yoga, where progress involves a two-way interaction between client and teacher, an essential condition for the growth of skills is that the client is in a state of active searching. Without this, little happens.

(A manual is being prepared for publication with instructions for teaching the Farmer and Blows method and containing guidance about instrumental applications.)

CLINICAL EXAMPLES

I would now like to present some clinical examples to illustrate the use of the relaxation and meditation techniques discussed here. Five examples with instrumental applications have been selected, one where the client used a method learned previously through yoga, three where the Farmer and Blows method was taught to the client and one where I taught CSM to the client as the method of choice for him.

Teeth-grinding

AF ground her teeth at night while asleep. She was well-practised at yoga. Perplexed about how I could help her, I asked her to relax, which she could do easily because of her yoga training. While she did that, the thought occurred to me that the action of closing the jaw could become a signal or conditioned stimulus to relax.

I therefore devised an exercise for her which she practised and she returned next session very happy with her results. She rewarded me with an account of her life history, which was very interesting but full of major forms of personal stress in Europe during World War II, which had in fact caused her symptom.

Following the initial treatment, she woke momentarily, relaxed quickly and resumed sleep. After a while the momentary waking ceased, as expected, without any grinding. This case is unusual because of the speed of the treatment. It is included to illustrate the advantage of yoga skills which greatly facilitated an instrumental application in this case.

Sexual perversion

AM was referred following conviction in a court of law for a sexual perversion. It transpired that in his particular case the perversion was not primarily a sexual disorder at all, but a secondary symptom of anxiety. It became manifest when he was over-wrought and very tense because of work and other demands on him.

He was taught the Farmer and Blows method and became very adept at detecting small signs of irritation throughout the day. Using the meditation response or 'relax' signal he was able to let go, forestalling reactions of anger and anxiety, as for example when his employees were slow to respond to his instructions. He experienced a growth in patience and overall enjoyment of life because of his creative use of the method.

Whiplash neck injury

BF had a whiplash injury from a motor vehicle accident. Subsequently her neck became a focal point of tension from various sources. She tended to experience strain doing physical work and pain during difficult situations with her husband's relatives, whose cultural expectations were quite different from hers.

These problems were discussed thoroughly and she was taught the Farmer and Blows method of relaxation. She enjoyed the general advantage of a break in the accumulated tension of the day and learned to relax her posture, avoiding her usual pains in particular situations such as bending over the kitchen sink, washing her hair, sitting at the hairdressers, and so on. She thus lost the symptom of pain which had been prolonged unduly after a physical injury.

Frigidity

CF, despite a generally good relationship with her husband, was unable to enjoy sexual intercourse with him. Because of over-rigid attitudes in her family of origin, she had never incorporated sexual responsiveness into her personality. She usually tensed on observing signs in her husband's behaviour that he might offer her more than a mere cuddle.

Her therapy programme included improvement of communication between her husband and herself and systematic desensitisation using the Farmer and Blows method, to lessen her habit of tensing on specific cues. Special emphasis was placed on her control over the conditions for touching or being touched and in administering the systematic desensitisation procedure. Experience of control in order to lessen elements of uncertainty and coercion is very important in most cases of specific anxiety, and may require a great deal of discussion in some cases.

After removal of CF's anxiety, which was tested out thoroughly during deep relaxation, response-stimulating steps were introduced. (This is another step-by-step procedure involving specific types of massage practised by the couple at home.) She was then able to enjoy full orgasmic release during love-making.

In some younger women, without a well-established pattern of not responding, the response-stimulating steps may be omitted because learning to 'let go' of interfering mental activity through meditation is sufficient. It is important in treating both men and women for sexual disorders to identify precisely any interfering anxiety responses.

Drug addiction

BM, a heroin addict, was brought to me by a friend, for his family had deserted him because of behaviour related to drug usage. Like many drug users, his ability to tolerate frustration was low and his personal discipline was minimal. CSM, rather than the Farmer and Blows method, was chosen for his use, following a lead from Carrington about the possible potential of meditation as a substitute for drugs of addiction.

BM described how heroin produced a state of relaxation in him, despite the side-effect of greatly reduced stress tolerance. I therefore helped him to use CSM as a healthy substitute, producing a similar effect of relaxation. According to the last report from his probation officer, who is responsible for his supervision following convictions at a court of law, he looked healthier than he had ever seen him before. The client continues to meditate twice a day and has not resumed heroin use. Meditation can thus offer help for at least some people with problems of addiction.

Migraine and hypertension

In parallel with work in Australia, there have been useful extensions of instrumental applications of relaxation methods at the Menninger Clinic in the USA, involving migraine and essential hypertension. Elmer Green's group of researchers have made use of the fact that when people relax deeply so that the parasympathetic division of the autonomic nervous system predominates over the sympathetic, the peripheries of the body, that is the feet and hands, tend to become warmer. This may be enhanced through the introduction of autosuggestion.[19]

Use may be made either of a combination of progressive relaxation, autosuggestion and biofeedback monitoring, as by the Elmer Green group, or the Farmer and Blows method and autosuggestion. A substantial number of migraine sufferers can make active use of these methods to avert a migraine attack, and the same process, practised long enough over a period of months, can reduce essential hypertension. Excessive menstrual contractions and other seemingly involuntary internal reactions may also be modified using relaxation methods, at least in a useful proportion of cases.

Stress management

Stress management for over-worked executives has become a popular topic. It is for those of us in this occupational bracket, which usually includes professionals, that stress management or relaxation 'packages' and 'mechanical techniques' are marketed. These quick and easy approaches are of only superficial help, though. It is better to learn a suitable method thoroughly. I have taught executives, using the Farmer and Blows method, to let go small amounts of tension during the day and to enjoy the advantage of a full break in mental activity at the end of a day's work.

However, such steps need to be placed in the context of an overall balanced orientation to living. Counselling is also necessary in most cases, in order to help reduce stimulation towards excessive striving, to establish a pattern of having a regular period of time out from work and to resist peer group pressure towards the 'willing horse' syndrome. As pointed out by Hugh Sedgwick in his book on stress,[20] it is important to place marital goals or other meaningful companionship above those of the 'career path'. In some cases a deeper inquiry is desirable into the dynamics of excessive striving, looking at such factors as fear of failure and ineffective defences against coming to terms with the condition of impermanence of such things as status and possessions. As is suggested by some of the Singer and Frankenheuser studies, the cases of factory workers could be worse than those of executives. In such cases, steps to permit the workers' control over the operation of machinery are very desirable.[21]

CONCLUSION

This chapter has been concerned with methods of 'letting go' so that excessive mental activity may be reduced. The connections with traditional Indian yoga have been described, commencing with the mantra methods of meditation of TM and CSM. With her development of CSM, Carrington helped to clarify the distinction between religious and psychological applications of meditation. The Farmer and Blows method originated from the Jacobson and Wolpe techniques of progressive relaxation, suspected of having connections with savasana. With the modifications, first by Farmer, then myself, this method has become more like the original savasana and so, in a sense, a circle has been completed.

Farmer's innovations increased the reliability for teaching and opened opportunities for many practical, instrumental applications. There could be an advantage in not using a traditional mantra for these purposes, although I do not see any particular difficulty in using a traditional mantra in those cases where the client understands and fully respects the religious origin.

Inherent in yoga there are many other potential adaptations. Some of these may be indicated briefly. A colleague of mine, Dr Sunder Das, uses pranayama (which is described in Chapter 3), and a technique of ablation which involves visualisation of a scene or a symbol which is gradually reduced in content until a state of stillness is reached.

Tarthang Tulku describes moving into negative feeling states during meditation, concentrating on the feeling itself while ceasing to think the thoughts that have been connected with it.[22] As well as being related to the approaches described here and to certain theoretical principles in experimental psychology, this approach is echoed by a number of other people, including Ron Farmer. He has recently developed a set of methods called 'nervous breakthrough' whereby anxiety may be seen as a harbinger of growth and opportunity rather than a signal for avoidance.[22]

Floating with anxiety is described by Claire Weekes[23] and by Pauline McKinnon.[24] All of these approaches are in line with earlier innovations by Dr Ainsley Meares, who developed ways of relieving the effects of stress by self-directed methods of relaxation without drugs.[25] Certain adaptations of yoga are on different lines, using symbolic representations of states of mind to be attained. There are many leads in this direction for the future.

With the disentanglement of relaxation and meditation methods from religion, it has become possible to integrate yoga with experimental clinical psychology and physiology. The inter-action between the two sets of discipline could be very fruitful for clinical practice. Just as the brain functions better with the integration of right and left hemisphere activity through medi-tation, so health care may be improved with an integration of the principles of experimental clinical psychology and traditional Indian practice. This has particular relevance for the correction of ailments in modern society, where people are subject to a rapid rate of change and are over-stimulated.

This conclusion might convey the impression that yoga is useful merely for what can be abstracted or adapted from it in the light of Western science. Such an impression would be illusory. In the higher states of consciousness, distinctions such as those made in this chapter between religion and scientific psychology give way to perceptions of unity, where distinctions of category and of purpose are transcended. It is yoga practice, irrespective of particular religious creeds, which leads to these higher states. The adapted 'letting go' methods described here may encourage some people to explore further.[26]

REFERENCES

1 Frank Campbell and George Singer, *Stress, Drugs and Health*, Pergamon, Sydney, 1983.
2 Fritjof Capra, *The Turning Point*, Wildwood House, London, 1982.
3 Patricia Carrington, *Freedom in Meditation*, Anchor, New York, 1977.
4 Byron P. Rigby, *Chairman's Address: Proceedings of the International Conference on Aviation*, Maharishi European Research, University Press Publication no. G1729, 1978.
5 H. Benson, *The Relaxation Response*, Collins, 1976.
6 Kenneth Pelletier, *Mind as Healer, Mind as Slayer*, Delta, New York, 1975.
7 Kenneth Pelletier, *Towards a Science of Consciousness*, Delta, New York, 1978.
8 Patricia Carrington, *Freedom in Meditation*, Wildwood House, London, 1982.
9 Michael Volin and Nancy Phelan, *Yoga Over Forty*, Sphere, London, 1965.
10 Richard Hittleman, *Yoga: The Eight Steps to Health and Peace*, Deerfield, New York, 1975.
11 M. V. Bhole, 'Concept of relaxation in shavasana', *Yoga Mimamsa*, 20, 1 and 2, 1981, pp. 50–6.
12 E. Jacobson, *Progressive Relaxation*, Chicago University Press, 1938.
13 J. Wolpe, *Psychotherapy by Reciprocal Inhibition*, Stanford University Press, 1958.
14 Ron Farmer, *Muscle Relaxation*, Self-Help Therapy Tapes, 149 Keen Street, Lismore, 2480, Australia, 1974.
15 J. H. Schultz and W. Luthe, *Autogenic Therapy* (six vols), Grune & Stratton, New York, 1969.
16 Patricia Carrington, *Freedom in Meditation*, Anchor, New York, 1977, Chapter 3.
17 Jagdish Dua, Personal Communication, Department of Psychology, University of Armidale, Australia, 1983.
18 M. L. Gharote, 'Importance of the hierarchy of yogic practices with reference to meditation and its results', *Yoga Mimamsa*, 21, 1 and 2, April and July, 1982, pp. 91–5.
19 Elmer Green, 'Self-regulation training: psychophysiological therapy in clinical practice', paper read at 7th International Transpersonal Conference, Bombay, 1982.
20 Hugh Sedgwick, *Stress and Counterstress*, Sun Books, Melbourne, 1983.
21 Frank Campbell and George Singer, *Stress, Drugs and Health*, Pergamon, Sydney, 1983.
22 Ron Farmer, *Nervous Breakthrough*, Breakthrough Therapy Tapes, 149 Keen Street, Lismore, 2480, Australia, 1980.

23 Claire Weekes, *Peace From Nervous Suffering*, Angus & Robertson, Sydney, 1972.
24 Pauline McKinnon, *In Stillness Conquer Fear*, Dover, Blackburn, Australia, 1983.
25 Ainsley Meares, *Relief Without Drugs*, Fontana, 1970.
26 M. W. Blows, *The Farmer and Blows Method of Relaxation*, in preparation.

PART III

The Parallels

9 *Yoga and Acupuncture*

HIROSHI MOTOYAMA

Western medicine has focused on an anatomical understanding of the human body. In this area of our knowledge the forms of the body, such as organs and other tissue, are well defined. However the human body actually consists largely of water in the form of body fluids. What about the bodily systems that function through this fluid medium without assuming any distinct physical form of their own?

In the newly conceived foetus, cells divide and multiply at an astonishing rate. Where does the energy for this amazing growth come from? The heart and circulatory system which will later distribute energy through the body have not yet come into existence. How is nourishment from the mother's body distributed throughout the child's body at this point?

Even at this earliest stage, the subtle energy system of the nadis, or meridians, is present. Even though this system has no shape or form like that of the veins and arteries that will follow, it is working like an underground water network, channelling vital substances to the developing body. Because the nadis, or meridians, lack discernible form, they are difficult to detect, and have long been overlooked by Western medical science. My own research and experimentation has led me to surmise that they function in the physical dimension through the medium of the flow of body fluids.

Both the disciplines of Indian yoga and Chinese acupuncture have recognised this subtle energy system, though they have called it by different names. I spent several years reading and comparing Sanskrit yoga texts and Chinese books on meridian theory before coming to the conclusion that these systems were essentially the same. Though the correspondence is not exact due

Table 9.1 Correspondence of the nadis and the meridians

Nadi	Meridian
Sushumna	Governor vessel meridian
Ida	Second line of urinary bladder meridian
Pingala	Second line of urinary bladder meridian
Gandhari	Third line of urinary bladder meridian
Hastijihva	First line of urinary bladder meridian
Pusha	Third line of urinary bladder meridian
Yashasvini	First line of urinary bladder meridian
Alambusa	Conception vessel meridian
Kuhu	Liver meridian (?)
Shankhini	Kidney meridian
Sarasvati	Spleen meridian (?)
Varuni	—
Payasvini	Gall bladder meridian
Shura	—
Visvodari	Stomach meridian
Saumya	—
Vajra	—
Citrini	—

to the different cultural contexts in which the systems were discovered and applied, I think the evidence is clear that they deal with the same fundamental reality. Table 9.1 shows the correspondences which I have identified between specific nadis and meridians, based on my reading of these ancient texts, on clinical experience and meditational insights, while Table 9.2 shows a tentative correspondence between the organ/tissue system, the chakras, the nadis and the meridians, based on descriptions in the Upanishads, traditional Chinese medical texts, clinical experience and meditational insight.

The basis of the discovery of the meridians and the nadis in China and India seems to have been the experience of masseurs, who noticed a series of flows while feeling for reactions in the body during massage, and of taoists and yogis, who recognised the channels of vital energy intuitively and extrasensorily during their meditation. Thus, similar energy systems were discovered and treatment systems developed accordingly. When these systems encountered each other, communication and mutual supplementation seems to have taken place easily due to their

Table 9.2 Relations between physical and subtle body systems

Organ/tissue	Chakra	Nadi	Meridian
Brain/nervous system	Ajna and Sahasrara	Sushumna, alambusa, ida and pingala	Governor vessel, conception vessel and urinary bladder
Respiratory system	Visuddhi	—	Lung and heart constrictor
Heart/circulatory system	Anahata	—	Heart and heart constrictor
Digestive system	Manipura	Kuhu, visvodari, payasvini, sarasvati	Spleen, stomach, large intestine, liver and gall bladder
Urogenital system	Muladhara, Swadhisthana	Shankhini, ida, pingala and kuhu	Kidney, urinary bladder, small intestine, liver and triple heater
Skeletal/muscle system	—	—	Liver, gall bladder and kidney

similarity. It was clarified recently, by historians Pierre Huard and Ming Wong in their book *Chinese Medicine*, that the nadi theory of yoga came into contact with the meridian theory through Nepal and Tibet about 2,500 years ago.

Two very significant theories of the meridians emerged in China at about that time, presented in the *Huang Ti Nei Ching* (The Yellow Emperor's Treatise On Internal Medicine). These ideas differed from the traditional dualism of yin-yang in that they divided all aspects of the universe into five parts and the body into three regions, each controlled by a separate energy system. These ideas seem to be essentially the same as the yogic concept of prana divided into five vayus (winds) and of the body divided into four regions. Although the number of divisions is different, the idea that the main part of the body is divided up, and that each part is individually controlled by prana or 'ki' ('Ch'i' in Chinese) energy is the same in both yoga and meridian theory.

One of the apparent differences between the Chinese and

Indian systems is their application. In acupuncture, knowledge about the flow of ki energy is used to identify which points along the meridians should be stimulated by means of inserting needles or burning a herb called moxa on the skin's surface. In yoga the flow of prana is regulated and the conditioning of the nadis effected by means of postures (asanas), breathing exercises (pranayama), meditation, and other such self-help practices. Actually, however, Chinese culture has its own traditional system of movement called Tai Ch'i which, like yoga, seeks to regulate the flow of subtle energy through conditioning of the channels. Acupuncture was a medical technique devised to treat those who were already suffering from some imbalance in their energy system.

Why is the conditioning of the meridians so important? It is because the meridians function as the intermediary between the three bodies of human beings: physical, emotional (astral) and spiritual. When energy is prevented from flowing smoothly, too much may accumulate in some parts of these bodies, and too little may be present in other parts. When such conditions persist, illness on all three levels will follow.

As an example, the spleen meridian is especially connected to the functioning of the stomach and spleen at the physical level. If energy flow is abnormal in this meridian, then malfunctioning of those organs will become evident. On the emotional level, abnormal functioning will result in obsessions, and on the spiritual level the person with abnormalities in the spleen meridian will be easily affected by others — possibly by troubled spirits, both living and dead. The proper stimulation of all the meridians will harmonise the functions of the three human bodies and produce a person who is healthy in the wholistic sense.

If a meridian is deficient or excessive in ki energy, then serious disease can develop. In my own clinical experience, sufferers of chronic diseases always show such deficiencies in one or more meridians. Over 10 years of treating sick people have convinced me that if the flow of energy in the meridians can be balanced, then any chronic disease can be cured, and the health of the whole body improved. We have cured even difficult diseases like rheumatism within six months to a year, and lesser complaints within a matter of weeks. Therefore I recommend to everyone that they practise some regular regime of exercise designed to

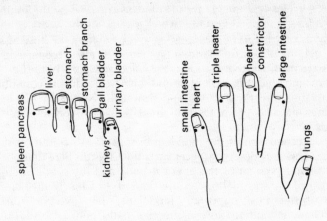

Figure 9.1   The 'well' points on the fingers and toes.

regulate their energy flow, before serious problems have the chance to arise.

I have formulated such an exercise programme for the use of my students and patients here in Japan, known as 'The meridian exercises'. If practised daily, it greatly facilitates the flow of ki (prana) in all the meridians. It consists of two parts. The first aims to stimulate and condition all the joints of the body. In yoga, these asanas are referred to as pawanmuktasana, or 'wind releasing' exercises; they are toe and ankle bending, ankle rotating and cranking, knee bending and cranking, hip rotating and stretching, hand clenching, wrist bending and rotating, elbow bending, and shoulder and neck rotating.

According to meridian theory, all the 12 major meridians course through the limbs, and they have their respective terminal, or 'well', points located next to the bottom corners of the nails (see Figure 9.1 showing the locations of the specific 'well' points). The 'well' points are very important, for it is here that ki energy enters and leaves the meridians. The energy level at these points is said to reflect accurately the condition of the entire meridian and, in the case of acute illness, acupuncture here is known to have an immediately beneficial effect.

In addition, many of the other important acupuncture points, where meridian function is reflected and where treatment may be most effectively applied, are also located in or near the joints. Therefore, pawanmuktasana, which moves all the joints of the body, is very helpful in stimulating and regulating the flow of ki throughout the entire meridian system. It is also recognised in Western medicine that the joints are vulnerable parts of the body, where fluid tends to accumulate and stagnate, leading to common complaints like rheumatism and neuralgia. Naturally, pawan-muktasana also promotes the circulation of blood and body fluids.

The second aim of the meridian exercises is to condition the spine. This area deserves special attention, for it is here that both the nervous system of the physical body and the subtle energy system of the spiritual bodies are centred. In Chinese medicine, one of the most important meridians, the governor vessel meridian, is said to flow along this path. Similarly, the all-important sushumna nadi, through which the tremendously powerful spiritual energy of Kundalini flows upward into the chakras upon awakening, is here.

It is written in the *Yoga Cudamani Upanishad* that the person who does not acknowledge the chakras in the body cannot become enlightened. This is one expression of the central spiritual fact that awakening of the chakras, those subtle energy plexuses along the sushumna nadi, is essential for spiritual development. Ki balancing and conditioning of the spine are a preparation for this awakening. Thus they not only improve the physical and mental condition of the aspirant, but also facilitate spiritual growth.

There are innumerable asanas for conditioning the spine, some described in ancient texts and some developed by modern teachers. I have divided them into four basic types: stretching; forward and backward bending; twisting; and asanas especially for the cervical area. If exercises from each of the four groups are practised, all the vertebrae will be comfortably moved and all the meridians stimulated. Many people suffer from slight to severe displacement of one or more vertebrae; this not only adversely affects their physical functioning, but results in imbalances in the flow of ki energy as well. Until these blockages are removed, Kundalini cannot properly rise. As individuals vary considerably, the selection of asanas for practice should be based on the needs of the individual body. However, as a guideline for students for general practice, I chose the following exercises in each of the four categories:

1   Stretching asanas – tadasana (upward stretch), hasta uthana-sana and pada hastasana (backward and forward stretches), and chakra asana (wheel pose, side stretch). Taken together, these asanas stimulate all the muscles and meridians in the body.

2   Forward and backward bending asanas – paschimottanasana (back stretching pose), forward bending in the full butterfly position (that is, seated with soles of feet brought together as close to the body as possible), pada prasarita paschimottana-sana (leg-spread variation of back stretching pose) and supta vajrasana (sleeping thunderbolt pose) variation where the hands are clasped and stretched upward above the head. Bhujangasana (cobra pose) and dhanurasana (bow pose). For the area between the lumbar and sacral bones shalabhasana (locust pose), leg raising as a counter-pose for the locust (lying on the back with the legs raised straight up together so that the lower back is pressed against the ground). These asanas are effective for correcting inward and outward projecting vertebrae, and for stimulating the meridians which course through the trunk and limbs.

3   Twisting asanas – ardha matsyendrasana (half spinal twist pose), or simplified variation for beginners. This asana is effective in correcting vertebrae which project to the left or right side of their normal positions. It also stimulates the meridians in the trunk.

4   Asanas for cervical vertebrae – inverted pose (second stage of the full headstand pose) and neck exercises. Although the full headstand pose is very effective, it can be dangerous if practised incorrectly, and the complete pose usually takes considerable time to master. Therefore, the inverted pose in which body weight is supported with head and feet are incorporated to stimulate cervical nerves and meridians. After this pose, the cervical subluxation or dislocation is corrected when the neck is bent front and back, left and right diagonally, and rotated slowly in a large circle.

The selection of asanas and the mode of practising them (sequence, number of repetitions, duration of poses, amount of stimulation, focal points, etc.) should be decided according to individual body needs. Therefore, proper understanding by the practitioner and guidance from a competent teacher are of real importance.

The methods of diagnosing meridian functioning are many. An experienced acupuncturist can do this by examining the patient's pulse, the appearance of the body and asking questions about the patient's daily habits. At my research institute in Tokyo, we have developed a machine called the AMI which diagnoses ki energy flow in conjunction with a computer. As these methods are rather technical and complicated, I will explain here a more practical method of diagnosis which can be used by untrained individuals.

It was mentioned previously that the 'well' points (see Figure 9.1) located near the bottom corners of the finger and toe nails will reflect the condition of the entire meridian. Thus pressing the 'well' points and observing the colour of the skin can yield useful information about the corresponding meridians.

Where the skin upon light pressure appears white or cold, ki energy in that meridian is deficient. Where there is warmth and a healthy red colour, the condition of the meridian is normal. Where there is pain on light pressure of the 'well' point, ki energy is excessive; and when pain occurs only as the result of deep pressure, ki energy may be seen to be deficient.

Similarly, the 'moon' of the nails on toes and fingers can show the state of the associated meridians. If there is no moon apparent, this is indicative of a chronic deficiency of ki energy flow. Likewise, excessively small nail size indicates a long-term deficiency in ki energy in the corresponding meridians. However a nail showing a normal pink colour indicates a healthy meridian.

Vertical wrinkles in certain nails shows a deficient blood supply to them, reflecting a deficient meridian function. Ridges or bumps in all the nails reflects such a deficient blood supply throughout the entire body, as does nail colour, which will be excessively pale or white. However, if only some of the nails show such lack of coloration, then the corresponding meridians will be deficient and in need of proper stimulation.

Another way to check meridian functioning is by noticing how the fingers perform when they are contracted into a fist and then opened out again. Fingers which feel or appear weak show weakness in the associated meridians. Deficient ki energy is shown by peeling of the skin near the 'well' points, and excessive energy is indicated when skin rashes occur at those points.

Yoga asanas, when practised properly, are most beneficial for stimulating the meridians and balancing ki energy flow. Regular practice of a balanced regime of exercise, with particular attention to the special needs of the individual body, will surely result in better health of the physical, emotional and spiritual bodies, and thus the harmonious functioning of the whole human being.

FURTHER READING

H. Motoyama, *Theories of the Chakras: Bridge to Higher Consciousness*, Theosophical Publishing House, Wheaton, Ill., 1982.

H. Motoyama, 'The meridian exercises', *International Association of Religious Psychology (IARP) Research for Religion and Parapsychology Journal*, 8, 1, Oct., 1982.

H. Motoyama, 'Yoga and oriental medicine', *International Association of Religious Psychology Research for Religion and Parapsychology Journal*, 5, 1, March, 1979.

10 *Autogenic Training in the Context of Yoga*

VELTA SNIKERE WILSON

Autogenic training (AT) is a specific method of relaxation. Developed by the German psychiatrist, Johannes H. Schultz, in the early 1930s, it has since spread widely. Apart from Germany, where it is frequently used as a routine pre-operative procedure, AT is popular in Canada, USA and Japan. Hannes Lindemann in his book, *Relieve Tension The Autogenic Way*, states that Indian psychiatrists practise autogenic training and not yoga. Will this chapter elicit verification or denial of this statement? Perhaps they practise both, like many other people.

WHAT IS AT?

The method was developed by observation and study of the beneficial side-effects of hypnosis and self-hypnosis. The benefits of relaxation are well known and not denied by anybody and I will take familiarity with them as granted – a switching from the dominance of the sympathetic branch of the autonomic nervous system, from the state of 'fight or flight' (to which I would like to add 'or rigid with fright'), to the parasympathetically-induced state of relaxation and regeneration. Once the parasympathetic system responds to conscious interaction and the recuperative state is available at will, then intended activity and unhindered unfolding of one's potential is also more readily available.

AT differs from self-hypnosis in that the suggestions given are not commands but invitations, and the body is left free to respond according to its need for the clearing of obstacles to optimum functioning. The attitude of the practitioner is that of passive observer, the 'activity' maintained being that of 'passive concentration'. One is active in so far as one makes the suggestions and maintains concentration gently restraining the mind from wandering, and passive in so far as one observes and receives the responses which proceed according to the self-regulatory processes of the organism.

I have known many methods of relaxation and will dwell on the ways in which I have found AT to differ from them. For a start, AT is the fastest method. Slowly and methodically, a carefully researched string of self-suggestive sentences is progressively learnt. Learning consists of passive attention to the reaction of the body. There is no imposition of will and frequently a spate of many differing reactions wells up before direct contact with the body is established. People differ in the time it takes for the pathways to be cleared. An established practitioner achieves the relaxed state in a few seconds or even instantaneously when words are no longer required and alterations in physiology happen at the point of internal directives. Once it is learnt, speed is one of the hallmarks of AT.

Another characteristic is that the body is not expected to react in a stereotyped way. Apart from the increasing speed of reaction, subtle changes constantly take place as the intensity of the practice grows. The suggestions are, moreover, changed to suit current needs as well as distant aims. Thus AT grows with the practice and the practitioner.

Before becoming familiar with AT, many years ago, I took part in a biofeedback course. During one of the sessions, suffering from a heavy cold, I relaxed very deeply and 'came back' without the cold. I quote this as an example of the well-known therapeutic potential of biofeedback, and the same potential can also be claimed by AT. Both myself and the people whom I have taught have experienced quite miraculous therapeutic effects brought about by the regular practice of AT. AT differs from biofeedback in that it needs no apparatus, can be practised anywhere in any circumstances and conditions and, of course, has the additional advantage of speed. In connection with AT, biofeedback is useful in ascertaining whether the felt state of relaxation is objectively demonstrable; several people who wondered whether the relaxation they felt was only a figment of their imagination had the machine demonstrate to them the fact of their ability to relax by quantifying it. The opposite, i.e. AT reinforcing biofeedback, has not happened in my experience.

PRACTICAL EXAMPLES

Autogenic training has terminated headaches, banished insomnia, helped in cases of asthma, migraine, agrophobia and general anxiety. It has changed outlooks on life and benefited relationships; it has liberated energy and increased efficiency.

For example, a slip of a young woman was brought to me clutching a hot water bottle for comfort. She was agrophobic and anorexic. A university graduate, she had scoured the medical field for available treatments over the previous 12 years, but without result. Within five weeks of learning AT, however, she had put on half a stone in weight and was going to parties. (A humorous footnote – human nature being what it is, she expressed no joyous surprise but just took it all for granted.)

Another example concerned a young journalist who suffered from a psychosomatically disturbed heart, to the degree that he felt threatened by imminent heart attack merely from the effort of turning round in bed. He spent his time between cardiologists and psychiatrists. At the end of the AT course he regarded himself as cured and was pronounced as such by the medical experts, who discharged him. Moreover, this man had learnt the technique well enough such that when he was subsequently smitten by sciatica, he cured it by the use of AT in three days.

Another man had been off work for three months because of an obsessive death phobia. On completion of the AT course he had no difficulty in returning to work and felt in harmony with himself.

As far as physical ailments are concerned, massive doses of AT can speed healing, almost miraculously; for example, a sprained ankle was cured in 4 hours. And provided one has enough self-discipline to practise AT when emotionally disturbed, changing one's outlook and attitude minimises the problem. But AT has to be slowly and methodically learnt and thereafter regularly practised.

AT AND YOGA

AT does not take much time, but does require self-discipline, a characteristic it shares with yoga. There is no dichotomy between AT and yoga, they blend rather well. AT is the threshold between the outer and inner limbs of the eight-fold path of raja yoga, and can be regarded as the initial step of pratyahara – the withdrawal of the mind from the external objects of the senses. It is a turning of the attention inward to one's own physiological processes. That is the first and basic step in AT.

Further, as in yoga practice, AT brings about purification – a general 'spring cleaning' of body, mind and emotions. Thus, in the learning of AT, long or even totally forgotten events surface into awareness. For example, one woman's hands became so hot she had to blow on them. Finally, she recalled having been told that her hands had been scalded when she was a baby. There were no scars and she had never before contemplated the event. Once the memory bubble burst, however, sensations returned to normal and her hands remained pleasantly warm. Another woman felt her cheek being gripped and pulled. By talking with her father, she discovered that the dog who attacked her when she was a baby had got hold of her cheek. Another person reported that for days her lips felt bashed and swollen. Frequently consulted, the mirror reported no such things. It took her a week to remember that as a toddler she had fallen on her face from some height. In other cases the causes of various strange sensations rooted in more recent events were more readily recognised, although rarely without some consideration and pondering. It seems as if memories of injuries, illnesses, scars not physically visible yet recorded somewhere in the psyche, are turned out like pockets and wiped clean.

People often drag about with them heavy suitcases, full of oceans of forgotten or unexpressed grief, that can spring open during AT so that sometimes currents of releasing tears are shed. In contrast, one man started laughing during a session and says he did not stop chuckling or smiling for weeks. It dawned on him that he had not laughed for five years.

When such changes take place and people ask 'Why do I feel so happy? Is something wrong?' my answer is 'Get used to it. Joy is our birthright.' Joy arises spontaneously when blockages to inner harmony are removed. As to energy being liberated and made available, practitioners have reported 'I can now do more things in less time.' Children and teenagers have found AT useful and an 11-year-old boy exclaimed, regarding AT, 'I wish I had done it years ago.' Teenagers who have become physically ill from the contemplation of impending examinations have, with the help of AT, sailed through them.

Apart from such physical and emotional memories arising from events in the current life, there have been cases of past-life memories welling up, with their meaning and resolution sometimes unlocking current-life problems. It has emerged from experience with AT that even suggestions given to oneself in past lives can surface. From my own AT experience, I understood that people about to be burnt at the stake impressed upon themselves visions of scenes of snow and ice and even birds and animals dying of cold. This vividly corroborated previously-recalled distant memories.

It is said in yoga that upon realising the fifth yama, aparigraha – non-greed, non-fixation – knowledge of past incarnations becomes available. Why such a claim is made has puzzled many. Teaching AT, I have noticed that people's arms getting cold rather than warm is an indication of a large stack of hoarded memories waiting to be released, more often than not from past lives. It is, of course, with our arms that we grab and hold on to, and this tenses them. If we fix ourselves to our belongings, ideas, beliefs, loves and hates, we are fixed in time and thus not free to flow with time. If we let go of tensions in our arms, if we let go of gripping of the subtlest modes at the subtlest levels, time looses its fixating grip on us. AT has thus helped to solve the puzzle of the consequences of aparigraha, elucidating the connection between aparigraha and its consequences.

The niyama tapas – heat, intensive practice – also provide links between yoga and AT. There is a general and very marked warming up of habitually 'cold' people; a sensation, difficult to believe at first, of a hot-water bottle having been placed on the back or shoulders or stomach. Electric blankets are discarded and persons who used to close windows and turn up the central heating do the opposite, sometimes to the alarm of the family as roles are reversed. Could one say this is a minute turn in the direction of the first step to be able to melt the snow around one?

Detachment is a well-known yogic practice. People doing AT report 'The circumstances that usually upset me so much rose again, but I could not be bothered to get upset.' No suppression of unwanted reaction, no effort, just 'could not be bothered to get upset'. Detachment arises spontaneously. Another initially baffling result of AT is an increase in bodily mobility and flexibility, but this can be easily understood as the letting go of an unconscious clutching at one's bones and joints with one's muscles.

WHO SHOULD TEACH AT?

We have seen that AT provides no contradiction to yoga; rather it facilitates its various practices and attitudes. As the Alexander technique is a helpful adjunct to the asanas, so AT greatly speeds up and facilitates relaxation and introversion. The question then arises 'Should yoga teachers also teach AT?' On the whole, AT is taught by medical doctors who insist that medical supervision is essential, on two counts. Due to increasing general awareness, one may also become aware, first, of subtle symptoms indicating incipient illness, symptoms that ought not to be glossed over and brushed aside as they may lead to early diagnosis and treatment. Doctors have reported such cases from their own personal experience. On the other hand, what medical circles call 'as-yet unknown self-regulatory processes' one could on many occasions call 'healing' in yoga terms, and this faculty may be more awake in, and more familiar to, people in yoga circles.

Equally, the AT instructor must be ready and able to deal with all matter of psychological effusions, so that a psychotherapeutic touch and technique are needed, a familiarity in dealing with psychological problems. That is the other point of argument for medical supervision. On the other hand, what would an orthodox doctor do with obvious, and possibly disturbing, reincarnational memories if he does not acknowledge their existence?

Hannes Lindemann certainly pleads for the delegating of AT teaching to a much wider cadre of people, even though most doctors insist that AT teachers should have experience in dealing with patients and should belong to the medical or paramedical professions. As AT is so beneficial and doctors have so little time even to speak to their patients, the time must be coming when well-trained, suitable people other than doctors will be accepted as AT teachers. Yoga teachers, because of the many parallels between yoga and AT and mutual facilitation, would seem to be next in line. Yoga practitioners learn AT more easily, the two disciplines blending and supplementing each other. I think that emotionally mature yoga teachers, well instructed, would therefore spread the benefits of AT and benefit yoga by introducing this streamlined and fast method of relaxation and introversion.

Minute description of the method can be found in the books listed. However, the attitude to be adopted, the role of passive awareness, more often than not has to be painstakingly explained and demonstrated again and again in many differing ways suited to the particular student. A book cannot make sure such instructions have been understood, and therefore a live instructor is valuable. My advice to anybody who wishes to obtain the great benefits following AT practice is – find an instructor, do the course, and only then read the books.

FURTHER READING

J. H. Schultz, *Das autogene Training*, Georg Thieme Verlag, Stuttgart, 1932.

W. Luthe (ed.), *Autogenic Therapy*, Vols. I–VI, Grune & Stratton, New York – London, 1969–70.

Karl Robert Rosa, *Autogenic Training*, Victor Gollancz, London.

Hannes Lindemann, *Relieve Tension The Autogenic Way*, Abelard Schumann, London.

11   *Yoga and Ayurveda*

STANLEY JACOBS

Let all be happy,
Let all be without disease,
Let all creatures perceive the good,
May no one be in misery of any kind.

> (A resolution uttered by sages
> on first rising in the morning)

Yoga has a natural affinity with ayurveda, for both of them have their particular existence and origin in the profound spiritual wisdom of ancient India.

WHAT IS AYURVEDA?

Ayurveda is called the science of life. It is essentially a wholistic medical system, and is experiencing a remarkable renaissance in modern India itself. Indeed, it has been said that ayurveda is the grandfather of all medical wholistic systems. As yet, however, it is little known in Western countries. As indicated by the World Health Organisation in 1980, ayurveda deserves to be more widely known. Its principles and comprehensive theories, instructive and profound as they are in their own right, could very well help to integrate complementary and modern orthodox medicine. In fact, its completeness is such that a modern scholar has said that even if all spiritual references were removed from ayurveda there would still remain a perfectly coherent system of scientific medicine.

Nevertheless, ayurveda is said to have been a gift, given by the Creator, in response to the overwhelming tide of disease and misery that began to sweep mankind some 5,000 years ago.

In medical terms, ayurveda has been classified into eight major divisions: medicine, surgery, gynaecology/obstetrics/paediatrics, toxicology, psychiatry, longevity, fertility and ENT. However, only a third of ayurveda is concerned with disease as such. It is mostly concerned with health and with providing a guide for life in general. It is thus an excellent system of preventative medicine and positive health. In this respect it comes close to the spirit of yoga, for ayurveda enjoins you to keep your body healthy, strong and willing to undertake vigorous work in the physical world. Similarly, it says that when you first wake up in the morning you should remember yourself as pure spirit, in whatever way you find most helpful. This brings a peaceful state of mind for the whole day. Next, think of your day's work in terms of service to others.

In ayurveda, the hygiene of the body includes such unfamiliar practices in the West as tongue cleaning to remove toxic mucus products; gum and eye massage to strengthen the muscles; and light oil in the ears to maintain their efficiency. Normally, men should shave and cut their hair and nails regularly, for these are said to be 'secondary products' of bone metabolism and through them many toxins are removed. Such practices are not so necessary for women, whose menstrual flow removes much of the toxicity.

Food should be fresh and of the best quality affordable. Generally, two meals a day are recommended. If possible, the first meal is best served around 11–12 am. It should be the largest of the day, since the gastric digestive fires are at their height at this time (in parallel with the strength of the sun). Breakfast, if taken, should be light, since the digestive fires are low in the early morning. Only a little water or liquid is recommended at meals, just sufficient to whet the appetite. If taken in larger quantitites it dampens the appetite, delays digestion, causes fatigue and can lead to obesity. Since the digestive powers naturally diminish towards evening, meals should be lighter than the main meal at noon and served not later than 8–9 pm.

Ayurveda lists 13 natural urges of the body, which should only be suppressed if there is an overriding need to do so, otherwise an imbalance of energy may arise leading to disease. These urges are urination, defaecation, flatus, vomiting, eructation, sneezing, yawning, coughing, crying, hunger, thirst, seminal discharge and deep breathing after exertion.

THE PRINCIPLES OF AYURVEDA

In ayurveda, the energy of the living body is called tridhatu or, less correctly, 'tridosha' which, when operating at the psychosomatic level, is equivalent to the term prana used in yoga. It is as well to remember that most of the energy we use comes through the atmosphere via breathing, and not from our food and drink as we often suppose. Hence the word prana as a synonym for breath. Generally, this energy differentiates into three different streams. This has led to a system of classification which relates the three streams of energy to a wide variety of phenomena, such as the five elemental powers of nature, the seasons, the times of day, food and medicines, pulse taking, types of pain and the tissues of the body, temperamental types, emotions and even to dreams.

The three streams are called kapha, pitta and vata (or vayu). Kaphic energy is especially concerned with growth and nutrition (anabolism); pitta energy with digestion (catabolism); and vata energy is related to the nervous system, the mind and to the chakra system of the subtle, vital body. Vata also controls and regulates the other two energies, both of which come under the general heading of metabolism. The central role given to vata (vayu) thus illustrates the psychosomatic nature of ayurveda and its essential link with yoga.

There is a fundamental relationship between the energy system and the five elemental powers of nature – space, air, fire, water and earth. Kapha energy is associated with the elements of earth and water; pitta energy with the element of light; and vata energy with air and space. The elements themselves are related to the five senses, to their five sense objects and to the five organs of action. Thus, space relates to hearing, sound and speech; air relates to touching, tactile sensation and to handling; light relates to seeing, colour/form and to walking; water relates to tasting, flavours and sexual activity; and, lastly, earth relates to smelling, aromas/scents and excretion. A little reflection will show how these sets are related together.

According to ayurveda, the abuse of any of these factors – elements, senses, objects and actions – by excessive use, underuse, or misuse is considered to be an important cause of unmeasured living and, eventually, of ill health. The alienation of modern man from his natural environment can be seen as the loss of the natural relation between the senses and the elemental powers of nature, at both the individual and the social level. In this way, modern man has truly 'taken leave of his senses'.

Here are a few examples to illustrate the kind of imbalances that frequently develop in our lives. Too much talking dulls the ability to listen, rapidly absorbs fine energy and creates an uncomfortable subtle space for the poor recipient of this excessive activity. Too much viewing, not only of TV but even of paintings in an art gallery, can dull the light of consciousness in the mind, tire the eyes, and cause one's obedient feet to ache. Wrong 'handling' of people, or 'rubbing them up the wrong way', can make them quite 'touchy' and they avoid making 'contact' with you. It is obvious that knowledge of these subtle interconnections, so common and popular in the vernacular, are intuitively familiar to us all. The reader may expand on these endlessly and it is particularly interesting to look at our own lives in this way.

So we need not only to come back to our senses but also to come back in the right measures. Knowledge of these measures is present in our heart. They come most readily when the mind is still and quiet, in the full awareness of the senses working in the moment. Then discrimination is seen to act by itself, the right measures become clear and the actions that are needed can be undertaken with confidence. Let us see how these connections can relate to quite a different area of experience – in this case the weather.

In rainy, spring weather, when there is active biological growth, the elements of water and earth, and the energy of kapha, predominate, whereas in the dry heat of a summer afternoon the element light, and the energy of pitta, predominates. When there is harsh, cold, windy weather, particularly in exposed areas like the moors in winter, the elements of air and space, and vata energy, are in the ascendant. Weather conditions such as these can predispose to ill health if precautionary measures are not taken, and can lead to so-called seasonal disorders. Thus in summer time, when overall heat and light are at their greatest, specific digestive power is naturally low since the body does not need to produce a great deal of heat or energy. If it did, a dangerous excess of 'fiery' energy would build up. So, in contrast to winter, hot summery weather brings out an appetite for lighter, cooler and uncooked foods, such as salads, fruits and vegetables. Cooked foods, 'rich' foods, sauces, foods strong in pungent, sour and salty flavours, and certainly alcohol, should be kept to a minimum. All these have either/or fiery and heavy qualities which need a fairly high digestive power to absorb. Ignoring these principles of eating – which we often do – leads to such common complaints as indigestion, sore throats, fevers, diarrhoea, etc. Ideally, one should modify diet and lifestyle a week or two before the actual change of season, although this may be difficult in countries like Britain which have such fickle weather.

When the energy balance is mildly vitiated then there are only vague symptoms of ill health, commonly known as being 'one degree under'. When vitiation of one or more of the energies is greater, then actual symptoms of disease develop, but without any 'objective' evidence at the physical level, as, for example, in the case of early arthritic pains, ulcer pains and frequent headaches. It is only when the energy is severely vitiated that disease enters into the very physical structure of the body. At this point, more specialised knowledge of medicine is needed to treat the disease. Thus, detoxification, or fasting, may become necessary, perhaps employing some of the yoga kriyas for body purification as well as the five chief methods of ayurveda – emesis, purgation, enemas, sweating and oleation. These elimination therapies are particularly effective for the more persistent and serious symptoms arising from disorders of excessive eating or undue dependence on junk food or chemicalised food.

The body, like the universe as a whole, is made up of a combination of the five natural elements. All bodies are but different combinations and proportions of these elements. They themselves, ultimately, arise out of the nature of the absolute substance, prakriti. All bodies thus differ only in form and proportion but not in substance. In human bodies, these elements combine in special ways to become transformed into seven tissue constituents called dhatus. These constituents are really subtle tissue essences. They are said to be the main sites for progressively refining the vitalising principle called ojas. Of the seven dhatu tissues, seminal fluid contains the greatest concentration of ojas. Hence, this is one reason why there is a widespread spiritual injunction for moderation in sex. Ojas also appears to be instrumental in the creation and quality of the body aura.

Part of the energy derived from ojas goes into maintaining our temperaments. There are positive and negative features in all types of temperament, except the seventh, when the person so endowed can adjust fully to any situation he or she may meet. Very briefly, a vata type can range from being either very scatter-brained or highly adaptable; a pitta type can be very efficient or extremely dogmatic; a kaphic type can be wonderfully loving and gentle or boorishly lazy. One point worth mentioning here with respect to vata types is that they should rarely be put on a short-term fast. Fasting enhances the air and space energies in the body and mind and as these will already be high, such a person may easily be precipitated into nervous and mental symptoms such as lack of coordination, breathing difficulties, dizziness and emotional instability.

The mind is also subject to the three gunas – sattva, rajas and tamas. These are said to be the all-pervasive forces or qualities of nature. Thus the mind needs to be in a pure (sattvic) state to experience unity, love, the presence of the real Self (Atman), truth, knowledge and discrimination. Precise, efficient action (rajas) is needed in order to be able to act from this state. And finally, the mind needs discipline (tamas) in order to hold the action fast to its true direction. The negative qualities of rajas and tamas form the essential factors that transform the power of love into the coarser bonds of attachment and identification. Generally, all kinds of psychotherapy and spiritual work aim at shifting the state of mind, i.e. the guna balance, from negative to positive, from less tamas to more sattva.

Having a realistic approach to life, ayurveda appreciates that sometimes a disease has gone so far as to require intervention with medicines, special therapies and/or surgical procedures. A prudent student of yoga would make use of this kind of help to forestall any serious decline in health which would interfere with spiritual work. In fact ayurvedic doctors use a variety of simple yoga exercises and breathing practices in all types of energy disturbances, as and when necessary.

THERAPY IN AYURVEDA

Therapy in ayurveda is divided into four basic categories – the spiritual, the psychological, the vital and the physical. We shall now consider this last category, the physical, in some detail.

The physical category includes yoga routines, as indicated

above, detoxification procedures, and often more specialised techniques. However, particularly central to ayurveda is the use of therapeutic substances which depend on the application of scientific knowledge and reason to the therapeutic properties of medicines, herbs and foodstuffs, extensively classified in a unique system called the rasa theory.

The rasa theory

This theory propounds that there are four ways in which therapeutic substances act in the human body.

Firstly, there is 'rasa' or taste itself. The effect includes both the subjective taste or flavour and the physiological effects produced by the substance while still acting in the mouth and tongue. Rather than list these latter effects, the reader is enjoined to go and suck a lemon to discover what these effects are in experience. There are six principal tastes given – sweet, sour and salty (sometimes known as the 'food tastes') and pungent, bitter and astringent (known as the 'medicinal tastes'). The following examples illustrate the use of the medicinal tastes.

Nature makes available those tastes which are most needed for every season. For example, during springtime, the desire to spring-clean arises naturally. This applies not only to the home but also to the body, where it comes from a natural desire to eat more salad foods. These particular tastes have a purifying effect on the body because they are particularly associated with air, which has moving, drying, clearing and contracting properties. On the other hand, the effects of sweet tastes are unctuous, filling and satisfying, because of their association with earth and water, and are more readily available in winter foods. However they are also commonly and legitimately used to relax people at the beginning of meals, although sweet tastes are the most abused, especially in their coarse forms.

Taste, therefore, is very important. The sense of taste may be objectively lost through illness, or subjectively lost in experience when, for example, one is too preoccupied or distracted during a conversation, when reading the papers, when listening to the radio – or even when attempting to engage in all three simultaneously! However, without the sense of taste there is a definite subconscious experience of loss and insufficiency. So unwittingly, we often compensate by filling up with excessive amounts of food and drink, not infrequently to 'bursting point'. Part of this sense of loss arises because we have not tasted the first few mouthfuls of food which, if it is good food, should always taste delicious. It is at this point the finest quality of the food is released – its sattva. And the experience of taste is important for other reasons too. It tells us whether the food is right or wrong for us individually; whether it is poisonous or wholesome; fresh or stale. Without this knowledge, unbalanced food intake must occur, eventually resulting in the incomplete digestion of food and leading to the toxic product called ama. Ama can eventually line the alimentary canal and can then seriously interfere with proper digestion. This is considered to be the single most important physiological factor in the causation of disease.

The second principal effect in rasa theory is called virya, or potency – the effect of the substance over the whole body while it is still in the mouth. The effect is either heating or cooling and there are ten sets of complementary qualities (gunas) intended to give information about the five elements. For instance a heating effect (ushna) can come from a high temperature or a fiery taste such as a chilli (a rather contradictory name in the English language, in view of its action).

The third principal effect is called veepaka. This is what happens when the food has been digested in the alimentary tract. It is often, but by no means always, the same as those of the first two. For example, cucumber is cooling in the mouth as virya, but after digestion as veepaka it has a fiery effect. Similarly, ginger, whose virya is fiery in the mouth, has a sweet veepaka.

The fourth rasa principal is the prabhawa. This is the unpredictable effect arising from the unique nature of the substance itself and relates to the specific active qualities of modern synthetic drugs. As we have hinted above, modern drugs and Western foods are beginning to be analysed by ayurvedic practioners in terms of the rasa theory.

The other three categories

The second therapeutic category in ayurveda operates through spiritual power like mantras, prayers and sacrifice, while the third category operates through the vital energy of the body – certain herbs, plants and gems are said to fall into this category. These two categories are a legacy from the most ancient Vedic period.

The fourth category of therapeutic agent is essentially the psychological practices related to the healing of mental disorders. In the past, although reassurance and counselling was given, it seems that the main emphasis of help for the more severely disturbed was some fairly rigorous down-to-earth applications reminiscent of practices in the last few centuries in the West. In this respect, one of the very promising developments of modern times is the evolution of new psychotherapies clearly inspired by the teachings in the Vedas, by yoga and by other Eastern spiritual practices. These integrate modern psychological and analytical understanding into quite powerful practices of psychotherapeutic healing and serve to act as a complement to the classical types. To a greater or lesser degree, the concept of Atman, the true Self, is the unique hallmark of these new therapies. By them people are enabled to recover their sense of inviolate transcendence and supremacy over affliction.

CONCLUSION

In conclusion, it is clear from this brief account of ayurveda and yoga that both disciplines have much in common. Both, therefore, can learn much from each other. Such a meeting can only help to quicken the spiritual development of mankind.

Let the final word on ayurveda be given to Charaka himself, the most famous physician of all.

This is named the Science of Life wherein are laid down the good and the bad life, the happy and the unhappy life; and what is wholesome and unwholesome to life itself.

FURTHER READING

P. Kutumbiah, *Ancient Indian Medicine*, Orient Longman, New Delhi, revised edition, 1969.

Pandit Shiv Sharma (ed.), *Realms of Ayurveda*, Arnold-Heinemann, 1979.

Bhagwan Dash, *Ayurvedic Cures for Common Diseases*, Hind Pocket Books, New Delhi, 1981.

12 *Biofeedback Meditation and the Awakened Mind: A New Gateway to an Ancient Path*

ANNA WISE

Yoga meditation is the ancient science of union of the self. Biofeedback is the modern science of electronic measurement of internal states. Biofeedback with meditation is a synthesis of old and new, a meeting of East and West that provides a unique way for Westerners to be introduced to the age-old traditions of yoga. Biofeedback is a new gateway to an ancient path.

Biofeedback meditation offers a bridge that can lead people from their ordinary waking state to the development of higher states of consciousness. As part of modern twentieth-century technology, it is attracting people to the study of meditation who otherwise would not find access to their own inner states of being.

Biofeedback itself is only an electronic technology, but in conjunction with relaxation and meditation training, it can provide a similar means of consciousness development as the practice of yoga. By using modern electrophysiological measurement methods to monitor yoga meditation techniques, it is possible in a very short time to teach people the methods of meditation best suited to their needs. Biofeedback is not an end in itself, but only a tool which aids individuals in the development of meditation skills. Once people have gained a certain degree of inner self-control and confidence, they no longer require the information or the aid of the technological instruments and tend to continue naturally in the practice of yoga, as we shall see. But first some background about biofeedback itself.

WHAT IS BIOFEEDBACK?

Biofeedback is the feeding back of your biology. In a Western medical sense there are many aspects of biology that can be measured and their physiological functions shown, or 'fed back', to the individual. There are machines that measure temperature, heart rate, muscle tension, stomach acidity and blood pressure, to name but a few. There are also many biofeedback devices in common use – ordinary bathroom scales are an immediate feedback of your weight; heart rate can be measured by anyone with a second-hand on his watch; a mirror aids you in adjusting your appearance. The biofeedback principle says that if, by some external measuring device or biofeedback instrument, you can be made to be aware of some physiological event of which you are normally unaware, then you can begin to learn to control it. Or as Dr Hiroshi Motoyama explains in *Science and the Evolution of Consciousness*, 'by making normal unconscious physiological processes conscious, one is able to alter subtle body processes at will'.[1]

At one time it was thought that the autonomic nervous system was the automatic nervous system – that is, body functions beyond our conscious control. Through the development of biofeedback machines, especially electrical skin resistance meters, people have been made aware of their own stress responses in the nervous system, and have been able to learn to control their own levels of arousal and relaxation. This is, of course, an ability that the practitioners of yoga have been aware of for centuries, without the need or benefit of external devices. However, many Westerners have not developed these abilities in the technological twentieth century and the practice of self-control of internal states was virtually unknown in modern America until very recent times.

When biofeedback was developed in the 1950s it grew up primarily in hospitals and clinical situations, and was used for pathological and disease related purposes – to control or alleviate migraine headaches, high blood pressure, ulcers, muscle tension and other stress-related disorders. C. Maxwell Cade, distinguished British physicist, psychobiologist and yogi, and the late director of the Institute for Psychobiological Research in London, introduced biofeedback into England with a different aim in mind. He maintained that, just as one can learn to control physiological functioning for a medical reason, one can also learn to control physiological functioning for the purpose of attaining higher states of consciousness. Some work had been done in America on alpha training in the hopes of reaching a meditation state, but Cade found this not enough. He wanted to be able to measure brainwaves at a number of different frequencies in both hemispheres simultaneously.

In order to develop this work, he invented, with Geoffrey Blundell of Audio Ltd, London, an electroencephalograph (EEG) that would measure the brainwaves of yogis in order to research and understand how they were producing their altered states of consciousness. The EEG, which they called the mind mirror, was able to measure both hemispheres of the brain at the same time and show the brainwave activity in the frequencies of beta, alpha, theta and delta.

1   Beta waves are emitted at a frequency of 14 to 26 cycles per second. It is the normal waking state of the brain, associated with logical thinking, concrete problem-solving and active external attention.
2   Alpha waves are emitted at a rate of 8 to 13 cycles per second. They are associated with relaxed, detached awareness, visualisation, daydreaming and a receptive mind. They can also be seen as a bridge from the subconscious to the conscious mind.
3   Theta waves are emitted at a rate of 4 to 7 cycles per second and are related to awareness of the subconscious, dreaming sleep, creative inspiration, peak experiences and deep meditation. It is from here that many of the spiritual 'ah-ha' experiences occur.
4   Delta waves are emitted at a speed of 0.5 to 3 cycles per second and are primarily associated with deep sleep. They can also be thought of as a kind of 'radar' of empathic, intuitive, unconscious waves.

Activity in the left hemisphere is generally related to logical, rational, mathematical, verbal and linear types of functions, while the right hemisphere tends toward emotional, artistic, musical and spatial functions. There is much proof, however, that the two hemispheres work more in conjunction with each other than was previously thought and the typical breakdown of right-brain and left-brain characteristics should be viewed mainly as metaphorical. Indeed the brain is much more complicated than we were once aware of.

Through his research, Cade found that the brainwaves produced by experienced meditators during meditation were a combination of alpha and theta in both hemispheres. The normal person in a waking state produces primarily beta waves in one or both hemispheres, while meditators in deep meditation were found to produce little or no beta waves. He also found that the yogis and swamis he measured had another fact in common: they tended to produce beta, alpha, theta and delta while in their normal waking states – in other words, they tended to produce some of the brainwave characteristics of meditation throughout their everyday lives. When they wanted to think or cogitate they just added the beta (thinking waves) to their continual meditation state. This state of consciousness he named 'the awakened mind'. In the 'awakened mind' state, the practitioner experienced not only the intuitive, unconscious 'radar' of delta waves, access to the subconscious knowledge and deeper understanding of the theta waves, but also the ability to process thought logically – and all at the same time!

Continuing his research, Cade began to find the 'awakened mind' pattern in other circumstances. He found that healers in the act of healing would often produce this pattern, as would inventors in the act of inventing, mathematicians at the moment of intricate mathematical solution, artists creating their masterpieces . . . all tended to use these maximum mental resources. Geoffrey Blundell, who co-developed the equipment and is also a director of the Institute for Psychobiological Research in London, explains further: 'Brainwaves have to be read in relation to each other, so that the meaning of a particular frequency on its own will subtly change meaning when other frequencies are also present.'[2]

Although the Biofeedback Society of America still maintains its focus in the direction of pathology and medical problems, a few of its members are also working on their own in the field of consciousness-training as it relates to physical and emotional well-being. It is now becoming increasingly acceptable to use meditation and Eastern relaxation techniques for physiological healing. One of the leading American biofeedback practitioners, Barbara Brown, talks about Eastern spiritual approaches to health versus the Western medical model:

> It has not been an easy union. Less than 20 years ago the potential of yogic practices for ensuring well-being was disdained by American scientific authority. Yet today, symposia, conferences and courses on the usefulness of yoga to recover and maintain both mental and physical health can be found in the curriculi of many of the most prominent American institutions serving the Health Sciences.[3]

The use of meditation and consciousness-training for emotional well-being is growing. Elmer and Alyce Green of the Menninger Institute state in *Beyond Biofeedback*:

> Every change in the physiological state is accompanied by an appropriate change in the mental-emotional state, conscious or unconscious, and conversely every change in the mental-emotional state, conscious or unconscious, is accompanied by an appropriate change in the physiological state.[4]

The Greens have conducted numerous research projects with Swami Rama that agree with Cade's findings about the value of accessing theta brainwaves, and have found that theta training is a way to open the door to a 'fabulous storehouse of memory'.

Pir Vilayat Inayat Khan, head of the Sufi Order in the West, in *Introducing Spirituality into Counseling and Therapy*, goes further in suggesting that in order to be receptive to the spiritual issues of the client, the therapist himself should perhaps be producing meditation brainwaves. Indeed, it is now beginning to be understood that individuals working in many different therapeutic circumstances tend to have a greater range of functioning in the alpha-theta-delta frequencies. This would help to explain their empathetic and intuitive abilities, as well as their sometimes innate healing powers. Khan also cites numerous experiments investigating 'the ability to transmit messages in an uncanny way', and finds generally in these experiments that 'it turns out that the receiver of the message is producing mostly alpha and theta brainwaves, which are often associated with meditative states'.[5]

After working with Maxwell Cade for eight years in London, I moved with my husband, acupuncturist and evolutionist Jym MacRitchie, back to my home in Boulder, Colorado, USA and in 1981 we opened the Evolving Institute. Our main focus is on the processes of personal and social evolution – researching, studying and developing new possibilities, opportunities and techniques. As part of our approach to helping individuals in their own personal evolution, a methodology we call 'the evolving process', we developed our biofeedback meditation training into 'the awakened mind programme'. We felt as Cade did, and expressed in his book about his work *The Awakened Mind*, that 'the aim is to expand as far as possible the range of states over which one has full conscious control'.[6]

Given that we now understand Cade's research into the 'awakened mind' as the pattern of greatest potential in brainwaves, we set about developing specific ways to train normal non-meditating Americans to begin to 'awaken their minds'. As Cade had already noted, and yoga practitioners had found long before him, we realised that it was much easier and more efficient to begin to train in higher states of consciousness with a very deeply relaxed body. Thus the first part of our training includes learning how to relax. To aid in this we use ESR (electrical skin resistance) meters that 'feed back' the arousal and relaxation of the nervous system.

The student begins by relaxing his body, usually in a sitting position, with eyes closed. Alpha waves are produced much more easily with the eyes closed rather than open, so it is recommended that beginning meditators start in this way. At the same time as learning relaxation, the student also begins to clear his mind, or as Pantanjali put it in his *Yoga Sutras*, to still 'the squirming of the worm in the brain' – in other words to reduce or eliminate beta waves. A typical guided meditation for learning to relax the body and still the mind is as follows.

RELAXATION AND MIND CLEARING

Allow your eyes to close
Begin to let your thoughts gently settle

Allow yourself to let go of any preconceptions and
  expectations
And gradually begin to relax . . .

If you have any unwanted thoughts or images, imagine
  them as tiny puffs of clouds that you gently blow away
  with your breath until the sky all around you is clear

Gently begin to withdraw yourself from your environment
Withdraw yourself from your surroundings
Withdraw yourself from your thoughts
Withdraw yourself into yourself
Into your own centre, your own serenity, your own peace
And relax . . .

Begin to focus on your breathing
Breathing easily and deeply
Breathing relaxation into your body when you inhale
And breathing away any tension when you exhale
Breathing relaxation into your mind when you inhale
Breathing away any thoughts
And relax . . .

Allow the muscles of your face to relax
Your forehead and the muscles around your eyes
The muscles behind your eyes
The muscles around your mouth
Your lips, tongue, throat and jaws
Are all deeply relaxed

Allow the relaxation to flow down into your neck and into
  your shoulders
Allowing the muscles of your shoulders to let go . . .
Let the relaxation flow down your arms all the way to
  your fingertips
Now allow the relaxation to flow through your chest and
  deep into your stomach
Take the relaxation right into the very centre of your
  body . . .
The very centre of your being
Allow the relaxation to spread across your back and down

your spine feeling the muscles of your back let go . . .
And now take the relaxation through your hips and pelvis
and down into your legs all the way to your feet and
toes
Relaxed and calm
Take a moment to check through your body and find any
remaining areas of tension
And allow them to let go . . .

Deep within yourself you can visualise and experience your
whole body as completely relaxed
Deep within yourself you can visualise and experience your
mind as quiet and still
Deep within yourself you can visualise and experience your
emotions as calm and clear
Deep within yourself you can visualise and experience your
spirit as peaceful
Deep within yourself you can visualise and experience your
body, mind, emotions and spirit in harmony

Many biofeedback practitioners all over the Western world use
similar relaxation techniques to allow their clients to begin to
calm their bodies and still their minds.

Left on their own after a relaxation, students might eventually
reach a depth where a true meditation or altered state of
consciousness would be experienced. However, many students of
these techniques also become bored or disillusioned with continuing the practice in this fast, goal-orientated society; they may
not often reach a state where they can truly experience the
transformative effects of higher states of consciousness. At this
point the students can be instructed in particular imagery or
guided fantasy that is specifically designed to access both alpha
and theta brainwaves and give them the experience they are
searching for. Here are some examples.

ALPHA-ACCESSING IMAGES

Begin to imagine or visualise all of the colours in a box of
paints. Not just the primary colours of red, yellow and
blue, but also the secondary colours of orange, green

and purple as well as their various tones, shades, and
tints . . . pink, light blue, olive green, crimson, gold,
aquamarine . . . and, of course, the opposites of black
and white.

In your mind reach out your hand and stroke the sleek fur
of a kitten.

Feel yourself standing under a hot shower; feel the
slipperiness of the soap on your skin.

Hear the bark of a dog at midnight, howling at the full
moon.

Hear the pealing of church bells on Sunday.

Hear the piercing sound of a police siren.

Hear the cry of a newborn baby.

Taste the sharp tartness of pure lemon juice.

Taste the smooth sweetness of thick honey.

Taste and smell freshly brewed coffee.

Smell gasoline fumes in a traffic jam.

Smell lavender and roses.

Experience yourself walking with bare feet on the hot
beach under the noonday sun.

Experience yourself receiving a soothing, easing massage.

THETA-ACCESSING IMAGES

The following are two examples of visualisations used for accessing theta brainwaves to be given after a relaxation.

Doors of perception

Visualise or create in your mind an environment. Imagine
an outdoor environment – somewhere pleasant for you.
It could be the seaside, the mountains, the country . . .

Try to see it very clearly.

Walk around it . . . experience the colours, the textures,
the scents, the sounds . . .

Now within this environment place a house.

Don't go inside yet; just see it from the outside.

Now we're going on a journey through this house.

So, going in through the front door and passing through
    an entrance hallway, a seemingly endless corridor with
    rows of doors on either side, and each of the doors is a
    different colour.
You walk down this hallway passing door after door after
    door.
Until eventually you come to a stop in front of a door on
    your right.
You notice the colour of the door; on the door you may
    find a sign or symbol – you notice this and understand.
And now open this door and go into the room which is
    behind it.
Take a few moments now to explore very thoroughly what
    is in this room – find out all about it . . .

(long pause)

You can make any changes that you wish in this room.
You have it in your power to transform, take away or add
    anything that you wish.
Take a few moments to complete anything that you want
    in this room.

(pause)

And now go back out into the hallway and continue your
    way on down the corridor – passing more and more
    rooms . . .
Until you stop at a door on your left.
You notice the colour of the door and whether there is a
    sign or a symbol.
And when you're ready, open this door and explore the
    room behind it.

(pause)

Make any changes you want to make in this room.

(pause)

And now, when you're ready, go back out into the

corridor and this time retrace your steps back down the
    hallway . . .
Back past the first door you entered . . .
Back into the room of mirrors . . .
Notice any changes there are in your reflection – your
    image.
Back through the room of mirrors . . .
Back through the entrance hallway . . .
Back out of the front door . . .
Back into your environment.
Make any changes you want to make in your environment.
Find a comfortable place to sit and to meditate.
Meditate on what you experienced inside the house of
    doors, and what the different rooms mean to you.

This meditation has specific images designed to access theta
waves and open the subconscious. Often very personal meaningful images appear in the rooms that may either bring information
and insight or require transformation or healing. It is an excellent
way to begin to clear up unfinished business in the subconscious
and programme or affirm creative changes that you wish to take
place. It can be repeated many times as there are an infinite
number of doors which can be opened.

Theta meditation 2

(after deep relaxation)

And as you move deeper into your meditation you have a
    sense of going down a well
Continue going *down* into the subconscious
Down
A feeling of moving downstairs . . . and through long
    corridors
Twisting and turning
Through archways and strange circular doorways
Always going down
Sometimes the grade is steep
Sometimes only a slight incline
But always going down . . .

As you move down through the tunnels and corridors you
    can see beautiful iridescent lights of all colours . . .

The atmosphere changes – sometimes hot, sometimes cool,
    sometimes damp, sometimes dry, sometimes misty,
    sometimes clear . . .

Frequently there are open doorways, passages and
    corridors that you could enter but decide not to

You see lights and hear sounds coming from these places,
    some inviting and some perhaps disturbing

But still you move on . . .

Something from very deep down beckoning you . . .

And now you see the end of your journey

The end of the path

There before you is the place to which you have been
    travelling

That mystical, magical place that is so important for you –
    you have kept going to get there against all odds

Find out now what this place is, why it is important for
    you, what there is to learn here and why you are drawn
    here

Find the resolution to what it is you are seeking

And take all the time you need to do this, beginning
    now . . .

THE AWAKENED MIND PROGRAMME

The Awakened Mind Programme is designed: first, to help the
student enter into a higher state of consciousness; second, to
teach him how to return there at will; third, to educate him in the
purposes of altered states; and finally, to stimulate him to further
development. Once a true higher state of consciousness is experienced, the student often wants to become a practitioner of
meditation. Whereas he may have originally started the practice
of biofeedback meditation for purposes only of stress management, for dealing with some medical problem such as migraine
headaches or ulcers, or for learning how to cope with erratic
emotions, he is now receiving those benefits plus the experience
of his subconscious unfolding and the development of his spiritual evolution. The use of technology attracts many people who
would otherwise never consider taking up the practice of yoga

but, once acquainted with the practices, students begin to acquire the interest, the discipline and the perseverance that are necessary for any students of yoga to develop their skills.

America is a goal-orientated society. Many people enter into the practice and study of meditation with specific goals in mind: to learn to be more creative; to solve personal problems; for medical reasons or for self-healing; to manage time better; or to further their spiritual development. Therefore, the next step in the Awakened Mind Programme is to teach students what they can actually do with these higher states of consciousness once they are attained. Meditations may include work with self-healing, mental fluency, personal growth, concentration abilities, problem solving, inner guidance, detachment and spiritual awakening.

Healing circle

The following is an example of a meditation that is used for self-healing. It is written specifically to access a deep theta state, along with alpha waves, because research has shown that the healing will be more effective the more relaxed the meditator is and the greater the amount of theta he is producing.

(beginning with deep relaxation)

There is almost a sense of moving backwards . . .
Backwards into softness . . .
Backwards into warmth . . .
Falling . . .
Falling quickly now, faster and faster into a deep trance
And the faster you fall the more relaxed you become
As you fall you are aware of passing through different
    levels of consciousness
They may be presented to you as images, sights, sounds,
    voices, body sensations, feelings, even smells and tastes
And you fall beyond them . . .
Now you begin to slow.
Still you are falling, but gently, much more slowly – almost
    drifting down
And down below you, far, far down below you, you see a

circle – and it is toward the exact centre of this circle that you are falling

A healing circle . . . getting closer and closer now . . . and the closer you get the more slowly you are falling

And you see that it is a healing circle, the very centre of which you are moving towards

Now hovering just above it . . .

And with exquisite gentleness and softness you come to rest in the exact centre of this circle . . . so that you are lying symmetrically in the centre of the healing circle which you can see or feel very clearly all around you

And the healing begins . . .

It may take many forms . . . you may feel strange and pleasurable sensations

And as the healing permeates your body, there may be messages that come to you or just an indication of where the next step on your path may lead you

(pause)

And now ever so gently you feel yourself rise out of your physical body into your spirit body, your higher self . . .

And the healing continues and becomes even stronger as it is directed to your spirit body

And it feels good . . .

(pause)

And now as your higher self begins to feel whole and full, you descend again

Passing through the emotional and mental bodies . . .

Cleansing them as you move through . . .

Back into the physical body once again

Taking into the physical body all of the healing you received in your higher self

Integrating that healing into the physical body

Feeling yourself aligned, integrated and in harmony within yourself

At peace

The benefits of the Programme

Students have had a variety of positive responses to the Awakened Mind Programme: better control of thoughts; better able to do two or more tasks at one time; improved visualisation; self-healing of emotional and physical problems; a more relaxed and stress-free daily life; more creativity; greater access to subconscious needs and inner guidance; greater knowledge and control of higher states of consciousness; increased ability to meditate; ability to 'produce better results, faster, with less work'; greater ability to control one's own life and destiny; more spiritual awareness and contact with the higher self leading to greater universal consciousness. Obviously, these benefits depend on the initial reason for beginning the Programme as well as the amount of dedication and discipline during it and the individual's own previous background and experience. Many people, however, find improvements and benefits in areas that they had not anticipated. Perhaps the greatest single reward of the Awakened Mind Programme is that it stimulates students to continue in the pursuit of a lifestyle that fully integrates meditation and yoga.

During deep meditation, when the practitioner is developing his skills at accessing theta waves, he often becomes aware of subconscious material which travels up to the conscious mind. Many psychological insights can thus be gained. This subconscious material can then be worked through therapeutically, to whatever depth and extent the therapist and the client choose; many hours of discussion can follow one deep revealing session of meditation, or there can be simply a quick acknowledgment of the experience and the practice can move on. My own preference is to carry out most of the necessary work in hypnotherapy form, allowing the client to create his own inner psychological changes by reprogramming and reinforcing himself within his own meditations. The discussions that follow are used primarily to 'ground' the client in the conscious understanding of what he has become aware of in his subconscious.

If the student is being trained individually, another aspect of biofeedback can also be used to develop the alpha and theta pattern of meditation or the beta-alpha-theta-delta pattern of the awakened mind. The therapist 'feeds back' to the student when his pattern is at an optimum; in other words, while the student is practising his meditation with the EEG attached, the therapist notes when he has optimum brainwave patterns and tells him, without disturbing his meditation. In this way the student learns in his subconscious mind what it feels like to be in a certain state of consciousness. He will then find it much easier to return to that state the next time he meditates. The use of modern technology can, therefore, greatly enhance and accelerate the learning process of the awakened mind.

In a group setting, it is much more difficult to spend any length of time processing each individual's subconscious experience. It is therefore necessary to allow time within the meditation itself for the members of the group to consider how their experiences relate to their everyday lives – how they will integrate and use them for their own benefits. It is then often therapeutically valuable for each group member to express briefly to the rest of the group what he has experienced and how it affected him. It is a matter of personal preference as to how much discussion follows.

In a group, where it is not feasible for every student to have a personal EEG, the students may undertake an individual diagnostic session before the beginning of a class. This is designed to look at their brainwave patterns in relationship to the optimum pattern of the awakened mind and tell them the major areas on which they need to focus in order to move more effectively in the direction of the pattern they wish to attain. During the group training each student, therefore, has an individual path to follow within the generalised meditation instruction. Often students also choose to have another diagnostic session at the end of the group meetings in order to measure their progress and set a new path for future meditation needs.

One of the features of this programme that attracts many participants is the fact that it is not dogmatic or sectarian. People participate from many religious, spiritual and philosophical backgrounds, and there is no pressure to develop or adhere to any particular dogma or spiritual technique or path. Many different types of meditation are introduced and students are encouraged to experiment and discover which methods or styles are most effective and illuminating for them. Our principles concur with Swami Janakananda Saraswati's statement in *Yoga, Tantra and Meditation*:

Originally Yoga and meditation were not part of any religious system, but had a broader objective: not to bind people in ignorance and prejudices but rather to transform and develop human consciousness, thereby creating free, independent individuals, who make their own experiences, fulfill their own potential, each living and functioning in a personal way.[7]

After students have practised from anywhere between six months to two years, a transition begins to occur. They become much less concerned with the accomplishment or end product of the meditation, and become more concerned with the process of meditation itself; in other words, they begin to shift from goal orientation to process orientation. At this time they are more involved in the true process of yoga. It is now that they are more receptive to the more traditional methods of meditation.

In advanced meditation classes at the Evolving Institute we teach a spectrum of traditional meditations, many stemming from the tantric practices, as well as deep theta meditations for developing awareness of and resolving unfinished subconscious material. The variety of meditations practised include kundalini meditation, yoga nidra, tratak and antar tratak, antar mouna, prana vidya, chidakasha, zen meditations, pure consciousness, chakra and energy balancing meditations, and advanced methods of self-healing. Some individuals now also begin to work with the practice of the transmission of energy for the purpose of healing.

After some time practising these meditations, the students may begin to understand detachment and recognition of the central self. There is a deeper experience and understanding of the universe and man's position in it. He begins to develop the fluidity and movement of consciousness, and the control and awareness of inner energy potentials, both for the benefit of himself and others. As Swami Satyananda Saraswati, founder of the Bihar School of Yoga, explains:

The culmination of meditation is self-realisation. When a person achieves self-realisation it means that he has contacted his central being and now identifies his existence, his life, from the viewpoint of the self and not from the standpoint of the ego. When he acts from the centre of his being, the body and mind operate almost as separate entities. The mind and body cease to be the real him; they are merely manifestations of the self, his true identity. So it can be seen that the aim of meditation is to explore the different regions of the mind and eventually to transcend the mind completely.[8]

At this point students rarely need or wish to use biofeedback equipment. They are familiar enough with their own internal higher states of consciousness to want to be released from the use of external devices. Certainly, they occasionally enjoy measuring their brainwave patterns, just to check themselves scientifically, but on the whole they are no longer students of biofeedback but students of meditation and yoga.

CONCLUSION

The use of sophisticated electronic biofeedback instruments provides many benefits through the measurement of internal states and can attract a section of the Western population which is not initially receptive to or has no awareness of the study of meditation and yoga. But what about the yoga practitioner who wants to study biofeedback? Basically, he has a head start. He can probably relax effectively, clear his mind and produce some alpha waves.

The more meditation is practised, the easier it becomes to produce and to maintain alpha rhythms, and the longer continuous alpha rhythm is maintained, the more often the individual experiences states of higher awareness.[9]

The experienced yoga practitioner may also be producing theta waves. So how would the yoga student benefit from biofeedback training? Through diagnostic procedures he would be able to find out scientifically the areas of his brainwave patterns upon which he could improve and be instructed in specific means to move more quickly to the optimum state he desires.

Again West meets East. And together the development of consciousness and the evolution of the human being unfolds.

REFERENCES

1 Hiroshi Motoyama, with Rande Brown, *Science and the Evolution of Consciousness*, Autumn Press, Brookline, Massachusetts, USA, 1978.
2 Geoffrey Blundell, *The Meaning of EEG*, Audio Publications, London, 1985.
3 Barbara B. Brown, *New Mind, New Body*, Harper & Row, London, 1974.
4 Elmer and Alyce Green, *Beyond Biofeedback*, Dell, New York, 1977.
5 Pir Vilayat Inayat Khan, *Introducing Spirituality into Counseling and Therapy*, Omega Press, Santa Fe, 1982.
6 C. Maxwell Cade and Nona Coxhead, *The Awakened Mind*, Wildwood House, London (Bookwise, Australia), 1979.
7 Swami Janakananda Saraswati, *Yoga, Tantra and Meditation*, Ballantine, New York, 1975.
8 Swami Satyananda Saraswati, *Meditations from the Tantras*, Bihar School of Yoga, Monghyr, India, 1983.
9 C. Maxwell Cade and Nona Coxhead, *The Awakened Mind*, Wildwood House, London (Bookwise, Australia), 1979.

13   *Yoga and Healing*

LILLA BEK

People find it hard to understand why, if healing is a natural force potentially available to all, some of us have the ability to heal while others do not. Why do some doctors, osteopaths, yoga teachers and therapists have that extra capacity, while others, no matter how highly trained, have not? What is this strange power to heal?

It is not possible here to deal with the subject of healing in any great depth. What we can say briefly is that on the physical level most people can achieve a certain degree of healing. By using their hands for some form of physical manipulation, they can usually help a person get more energy into their body, while on a mental level they can give some support and confidence; they can encourage them, urge them on. However these are obviously not the highest healings, which are often silent; being able to heal anyone and deal with any kind of energy is essentially a case of spiritual advancement. Thus unless we are a 'natural', unless we have healed and worked with these vibrations in past lives, we will need to train ourselves, we will need courses and yoga in order to develop our intuitive abilities. Looking at this another way, if a person does not have enough of the blue ray – indeed, if he is lacking in any ray – this is one of the reasons that he has been reincarnated. He has come to work with and develop those areas.

To a certain degree, everybody *does* have the potential for everything; for example, if we massage or press certain parts of our bodies we can greatly improve our voices, even though we may not make operatic material. We have to realise, though, that behind anyone who has achieved anything, there is an accumulation of experience. So, the more effective and gifted a healer, the more likely it is that he has developed that knowledge in previous incarnations. The ease with which he can heal (and here we are talking about the laying on of hands) will depend on the quality of energy he can channel through his body, and that in turn will depend on the knowledge he has managed to accrue from past lives.

VIBRATION AND HEALING

To be able to heal in any real sense we have to have the blue vibration in our auras which enables subtle changes to take place on etheric levels. With not enough of this ray it is not possible to give any lasting help. The whole point of evolution is to quicken our vibrations; our chakra systems and our energies thus depend on how evolved we are. If we have big chakras, big energy fields, we will be able to heal people with equally large energies, but it is difficult to heal people with larger energies than ourselves. In the same way, it is difficult to tune in and heal someone who vibrates faster than ourselves. If there are rays and colours we are unable to produce and our patient actually needs those rays, we will be unable to help. Of course, there are times when a patient simply puts up a wall of fear, a barrier of non-recognition, and then the healer or therapist will not be able to reach him.

So what can we do to develop our healing capacities – what can we do to develop the blue ray? First of all, we should go deeply into the whole subject of healing. Doctors, after all, are required to dedicate themselves to many years of training, as is anyone entering the diplomatic corps. Healers require the dedication of both doctors and diplomats, for, whether they know it or not, they are ambassadors for the healing art. If people ask them questions they cannot answer, they will not be fulfilling their duty. It is important, therefore, to know the history of healing, to know the history of the body itself, to know some anatomy, to understand the continuity of life after life, to understand what counselling means and how to bring information through. We need to know what we are doing and what can happen. We need to know how to open our chakras and how to protect ourselves. Trying to open before we are ready can be disastrous for the physical body and if we try to give healing without relaxing, while we are in a state of tension, we can ruin our health, and even bring on a heart attack.

YOGA

One aspect of healing is the ability to create a positive environment, whether through treatment, attitude, relaxation, meditation, exercise or vibration, so that the patient can heal himself. This, of course, is what yoga aims to do. Yoga trains us to create a positive environment in which we can heal ourselves.

Yogic training is obviously helpful for relaxation and for control of breath and heart. It must be stressed, however, that yogic training does not simply mean doing a few exercises, a few breaths; it means a commitment and a deep understanding. Such training will then lead us to be more sensitive, to be more aware of what our chakras are doing and what is happening. Putting the hands in different positions, for example, can trigger off slightly different energies and it is important to tune in and be aware of this. We can usually tell when a chakra is speeding up or getting larger by a feeling of expansion or heat, paralleling the feeling of sexual climax. One chakra speeds another up to a point where the speed goes over a certain velocity and they all become lighter; there is a heady feeling due to a change of brainwave and we may feel light and slightly giddy.

My own job with the Federation of Spiritual Healers is to look at chakras and tell people about the state of their energies. Any yoga teacher should be able to tell you whether or not your chakra system is ready for healing, but even if there is no one to help you, you can usually tell when you are ready to develop. Firstly, there should be no sense of overcharging. You should feel relaxed and start getting indications in your life. When the time is right, the teacher usually will be there. In most healers' lives that I have seen, they have been directed in some beautiful way to somebody who has helped them fulfil their abilities. Some person, or perhaps the right book, the right idea, the right word, might have appeared and everything fell into place.

Yoga is a practice which should lead naturally into healing, just as healing should lead to an understanding of movement. At present, though, yoga practitioners tend to be more interested in self-development rather than healing, while many healers tend to ignore, or not bother too much, with their physical bodies. There is an irony here: the reason that so many people need healing is that they are moving so badly. Often certain areas are unable to clear because the arms never go above a certain level; people simply never stretch. This is where yoga asanas are so helpful, for cleansing, for exercising muscles, for stretching and coordinating the body, for changing consciousness. It is, after all, pointless going to an osteopath and spending vast sums of money if a simple posture can align the body and put your neck back into place. So behind every healer and therapist, there should be a knowledge of movement, together with a knowledge of breathing and meditation.

YOGA AND COLOUR

By practising yoga asanas we should understand, too, that we can activate particular colour ranges. In other words, yoga asanas are natural colour movements; in a sense, practising certain asanas is similar to colour therapy. Our chakras vibrate at different speeds and these are the vibratory rates used by colour therapists; we can shine a colour on a person with a lamp, project it to him with the mind, advise him to wear certain colours or use them in the home and suggest that he practise certain movements. There would, of course, be a problem if a person could not do a particular asana. For example, navasana (boat pose) is excellent for stimulating the muladhara, the base chakra, but for some people this asana is impossible. They therefore have to absorb the colour in different ways, through light or through healing.

When energy is lacking on the physical levels, warm colours are needed. The mental level requires yellow, the intuitive blue, the creative violet – although, of course, there is no set rule for any human being, and when using colours we have to be highly intuitive. If we are unable to *see* the centres, then we must sense them. A simple way of monitoring and telling whether or not centres need boosting is to place the hands on either side of the body and run them down the back and front. You can then tell if there is too much agitation in any centre, if a chakra is over-speeding or too lethargic. In the beginning, these two extremes are probably all we can hope to detect. A good idea is to work with coloured ribbons and see how the hands react. Eventually, as we go up the chakras, there should be a feeling like playing scales on the piano, a lovely sense of awareness as each chakra rises to a slightly faster level.

The question is sometimes raised whether any particular

conditions benefit from healing or colour therapy (here healing implies treatment that involves no physical adjustment). Perhaps it is easier to turn this question round and discuss conditions that healers are notoriously *bad* at treating. It is difficult, for instance, to find healers who are good with acute appendicitis, broken bones or teeth. Similarly, few healers can get rid of bunions or callouses. Such conditions respond better to orthodox physical treatment. Often there are sound reasons for attending to, or rearranging the physical. Perhaps the patient is walking, sitting or standing wrong, or perhaps he is wearing tight shoes and, because of the pressure, develops pain in other parts of the body; if he is given colour or spiritual healing, without any notion that his shoes are too tight, he will have to return for healing again and again because he is continuing to do the wrong thing. In the same way, if the patient's neck is out of place he will continue to have headaches until it is back in alignment, and if he fails to express his creative side no amount of healing will work until he is able to release this.

Some people may thus need to change their diets, take herbs or purify an organ, or realign their bodies. Some people are especially physically orientated and unless they swallow something, unless they are massaged or manipulated in some way, they will not believe anything is happening. Colour healing, aromatherapy, yogic movement and laying on of hands are all, no doubt, healing to a high degree, but if a person's body is lacking in tissue salts or vitamin C and he fails to take these, he will always be leaning on the healer or therapist.

On the whole, the people who perhaps benefit the most from these types of healing are the more advanced souls whose illnesses cannot be cured by ordinary medicine. If such people do take the prescribed drugs, the side effects are often as bad, if not worse, than the illness itself.

CONCLUSION

In the end, the success of any healing lies in the capacity of the healer himself. One who naturally has a lot of red energy will be good at dealing with arthritis, sciatica and rheumatic conditions, while one who has a lot of green will be good for cancer, and so on. Any healer may achieve a certain standard, yet specialise in a certain ray, which will draw people with particular conditions to him.

No doubt there is a change in the general attitude towards illness and treatment. Perhaps the biggest change of all will come when the drug companies start to explore and invest money in new areas of research. In the meantime, allopathic medication will probably continue to be administered, although hopefully in less drastic and perhaps even in more innovative ways. For example, there is a Dutch healer who projects the prescribed remedy with flashlights; he puts the medicine between two celluloid pads, lights a lamp and the patient absorbs it through the skin. Perhaps the breakthrough may come when orthodox medicine turns to more preventative measures. There could be systems of monitoring whereby any negative changes of cell structure or disturbance in organ rhythm would show up at an early stage. People would then be able to take appropriate measures before the illness gained hold. Clearly it is easier to get rid of a small growth than an enormous tumour which has pushed the patient to the edge of death.

What most people really need is more relaxation and fewer pills. Here, television and the media could play an active part, with programmes and videos of therapists taking people into relaxed states. Ultimately, the day may come when therapists and doctors may meet. Doctors will take into their curriculum the knowledge of the chakra system, with all its subtleties and implications, and therapists will be as highly qualified and as authoritative on medical matters as doctors. Until that day comes, we must each of us do what we can to develop our own innate abilities to help alleviate the suffering that exists all around us.

14 *Yoga and Natural Childbirth*

JANE SIMONS

The creation of new life is one of the most natural human acts. Like many other human experiences, however, it has been 'improved' on and interfered with until it has become totally unnatural – perhaps as unnatural as so many other aspects of our lives. However, a growing awareness of the need for a natural approach to health has recently influenced attitudes to childbirth. If we are to live as 'an organic part of the biological and cosmic universe and subject to the unchangeable and irrevocable laws of nature',[1] then we have to acknowledge that when it comes to entering this world, Mother Nature often knows best how we should go about it.

Both partners need to prepare, not only for the day of the birth but also for the nine months of pregnancy ahead. Being prepared to make the best of pregnancy will build confidence and self-reliance, influencing the woman's management of labour and motherhood to follow.[2] Her self-reliance in labour will be dependent on several factors: an understanding of the normal process of birth; the physical preparation of her body through diet, exercise, breathing and relaxation; and being at ease in the environment in which she has chosen to give birth, together with trust in the people who will support her at that time.

PHYSICAL PREPARATION

The mother-to-be learns from the very beginning of pregnancy to cope with aches, pains and discomforts as her body goes through various changes in chemical and postural balance. Since feelings of comfort or discomfort in pregnancy can make a difference to the relationship between mother and baby, a secure attachment bond is more likely if the mother is comfortable with her pregnancy. Therefore, a carefully selected programme of breathing, exercise and relaxation can help to counteract the softening effects on her spinal support, help in the support of her pelvic organs and help her to deal with backache. Yoga is very effective in teaching the pregnant woman to use her energy productively, to help her maintain a shifting sense of balance and to keep her joints and muscles supple and flexible so that she can manoeuvre easily from one position to another. Yoga helps to increase awareness of correct posture and to make best possible use of the supportive functions of the back and abdominal muscles, so counteracting the effect of diminishing muscle tone that usually occurs in pregnancy.

This increased awareness also helps her to preserve her back. By being aware of where her centre of gravity now falls, she is more conscious of how she lifts. And, should she suffer from strain as a result of lifting incorrectly, she needs to know which movements and exercises to practise to ease the resulting pain. This is important even after the pregnancy is over, when she again has to learn to use her body in a different way to adjust to the 'normal' effects of gravity.[3]

The pregnant woman also needs to be aware of the function of the pelvic floor or perineal muscles, for during birth it is possible to tighten these muscles unknowingly in response to the pressure of the baby's head. It is therefore important to exercise these muscles, not only so that she will be able to relax them during the actual birth, but so that she can also overcome some of the symptoms of weakness during pregnancy, like loss of urine when coughing or sneezing, haemorrhoids, or simply a feeling of engorgement when she has been upright for long periods of time.

One of the most important aspects of pre-natal training is relaxation and this goes hand-in-hand with training in breath control. Grantly Dick-Read and Lamaze were two of the obstetricians instrumental in the movement towards the inclusion of breathing techniques and relaxation to help the mother in labour.[4] Training in breath control evolved out of the need for women in labour to learn not to hyperventilate or hold on to their breath. Both are done as a natural reaction to pain or stress in any circumstance, not only in labour, and slow, deep breathing helps the mother-to-be relax and regain control of the situation.

Of course, not all women behave the same way in labour. Some find themselves panting in short breaths at the height of their contractions. This is where pranayama (breath control) techniques of yoga prove most useful, because they teach a variety of ways of using breath and so increase flexibility and adaptability in combining many kinds of rhythmical patterns. Some women find that moaning and groaning is the natural response to pain in labour – a perfectly rhythmical pattern in its own right and a wonderfully effective way of letting go.

Many women find that slow, deep breathing induces a feeling of relaxation. Concentration of the flow of respiration through the nostrils and the use of some simple techniques of yoga nidra may divert attention away from the source of pain. Breathing into the stretch of tight muscles gives awareness of pain management. Submitting with a long exhalation breath is a common response to sustained pain, especially a pain that is familiar. Many childbirth educators also teach how to massage pain spots in labour, giving couples the opportunity to support each other through non-verbal communication by touching and stroking. Using relaxation techniques takes some practice, however, and the earlier in pregnancy these are learned, the easier it becomes to deal with symptoms of stress and prevent them from occurring in the first place. All of this makes coping with labour and motherhood a much easier task.[5]

Although the increasing intervention of medical science in birth has given us the impression that there is only one position from which a woman can deliver a baby, there are in fact many different positions. Practising these different 'postures' in pre-natal classes gives women the confidence to experiment when the time comes, so reducing her fear of the unknown and strengthening her ability to manage the labour successfully. These postures range from simply walking about to 'belly dancing' while standing, or on all fours, or even squatting.

Women who are already adepts in yoga by the time they become pregnant often find themselves in yoga-like postures when they give birth. Some students have reported, for instance, that they gave birth in the 'cat pose', while others found it more comfortable to deliver from 'the armchair' or the cross between the 'crow-walk' and a crawl.

FAMILIARISATION

One of the advantages of joining classes early in pregnancy is the regular contact with other pregnant women. This establishes a sense of identity and belonging, banishing her fears and clearing up misunderstandings. And as couples tend to migrate from place to place these days, sometimes even from country to country, so women find themselves isolated in new working environments and unfamiliar neighbourhoods. Classes can then be an important source of support.

Another important part of the preparatory process is familiarisation with the place where the woman will give birth. The style of delivery chosen will depend on the facilities each hospital provides, and in many major towns and cities throughout the world these facilities are being updated. Although labour wards are still being run both to cope with the large turnover of patients and to suit the schedule of the hospital, most health authorities now take a more humane attitude towards the birth process. For example, a generation ago it was unheard of even to have husbands present at the birth. Some hospitals now have what they call 'delivery suites' or 'birthing centres' in which an attempt has been made to create a more homely atmosphere, with softer lighting and less evidence of clinical medical technology.

Some of these centres have been established as a reaction to the trend for couples to have their babies at home. A growing awareness of the influence clinical surroundings have on the birth of a baby is at the root of this trend for home births.

A HISTORICAL PERSPECTIVE

A few generations ago, home birth was the norm. Like more traditional societies today,[6] birth was purely a woman's business. A woman in labour was supported and cared for by the women in her family, assisted by a midwife. However, although she had the security and reassurance of familiar surroundings, birth was a risky business. Many women died of post-partem haemorrhages or pre-eclamptic fits, and the hospitals provided little in the way of a safer alternative, with their risks of puerperal fever and other infections. Medical science did gradually eliminate these risks but in the process made childbirth 'scientific', taking it *entirely* out of the control of the mother. Women abdicated all responsibility, relinquishing most of their natural ability to cope with the birth in the process.

In recent years, though, women have become more conscious of their individuality and have taken back the responsibility for their health and well-being. The movement towards natural chidbirth in safe but pleasant surroundings, together with the introduction of breathing and relaxation techniques, has altered the whole experience of birth. Education on the childbirth process for both men and women, together with a more enlightened attitude among health professionals, has prepared many women for a positive, fulfilling experience.[7]

Yoga, with its emphasis on self-awareness and self-reliance helps to develop the necessary skills to prepare a woman for one of the most important and wonderful experiences of her life. Giving birth has become a natural experience that women can look forward to with confidence and trust in their own natural ability to manage the birth of their babies – as naturally as they manage the rest of their lives.

REFERENCES

1 Paavo Airola, *How To Get Well*, Health Plus, Arizona, 1974.
2 Elizabeth Noble, *Childbirth With Insight*, Houghton Miflin, Boston, Massachusetts, 1981.
3 Elizabeth Noble, *Essential Exercises for the Childbearing Year*, Houghton Miflin, Boston, Massachusetts, 1976.
4 Grantly Dick-Read, *Birth Without Fear*, Pan, London, 1969.
5 Michel Odent, *Birth Reborn, What Birth Can and Should Be*, Souvenir, London, 1984.
6 Frederick Leboyer, *Birth Without Violence*, Fontana, London, 1977.
7 Sheila Kitzinger, *Pregnancy and Childbirth*, Doubleday, New York, 1980.

15 Yoga and Osteopathy: Health Care for the Year 2000

ROMAN A. MASLAK

Osteopathy is a practice of medicine involving the physical treatment of the body and spine in order to remedy disease. Specifically, it is an independent medical therapy which focuses on treating the cause of disease and not just the symptoms. Its role is in primary-contact general practice and it deals with a wide range of patients and problems, regardless of age or complexity. It is thus not merely limited to manipulation. It shares with yoga and other alternative therapies a wholistic philosophy, embodying the principle that any treatment programme should acknowledge the relationship between mind, body and spirit – that is, the total being.

On first inspection one may be forgiven for seeing little or no similarity between osteopathy and yoga; from a medical perspective, apart from their mutual interest in the body, each is distinctly different. However, functionally they share many common elements which are particularly accessible to yoga practitioners. Before we can understand how these two systems relate, however, let us first look at how osteopathy works and how it developed.

WHAT IS OSTEOPATHY?

It is 2,000 years since Hippocrates, the 'father of medicine', advocated that physicians turn their attention to the patient and not the disease. Ill health, he said, was not an affliction thrust upon man by jealous gods or external sources; it was the result of natural causes. The role of the physician was to ascertain what had gone wrong in the patient's nature and then to proceed to aid nature in its recovery, rather than getting in nature's way. The practice of osteopathy, developed by its founder, American mid-western physician Dr Andrew Taylor Still, is based on this ancient principle.

In his day, Still saw how medicine was rapidly moving away from Hippocrates' principles. Medicines were abused and surgery was over-zealous; indeed the Cnidian misconception which dictated extracting the disease rather than restoring health was becoming the rule (and how true this is today). Still felt there was a better way of treating patients. He believed that imbalances and insufficiencies of the circulatory and nervous systems rendered the body vulnerable to disease (our present dilemma with stress and heart disease being a prime example). Correction of these imbalances could be found in the musculoskeletal system, which comprises more than 60 per cent of the body's bulk. To suggest that no relationship existed between systemic function and skeletal integrity would be a gross mistake.

Still's research revolved around a painstakingly detailed study of the musculoskeletal system in health and disease, noting subtle changes which took place in the soft tissues, bones and joints. These differences, which were frequently restrictions in the muscles, fascia and joints, became known as 'osteopathic lesions'. Specifically, an osteopathic lesion is any structural abnormality (excluding trauma or birth deformities) that leads to functional or organic disease. From this appreciation of body dynamics came Still's tenet 'Structure governs function.' His solution was not medicine or the knife, but a 'soft hand and gentle movement' to restore musculoskeletal function. And in 1874, osteopathic manipulative therapy was born.

Considerable research has taken place into the physiological basis of the osteopathic lesion and anyone interested in a detailed technical explanation should consult the works of American physiologist Dr Irvin Korr[1] and the journals of the American Osteopathic Association. For our purpose here, however, let it be said that osteopathic treatment is concerned with pathophysiological reflexes which can modify the intricate balance of the autonomic nervous system. Those functions of the body not under voluntary control are influenced by the sympathetic and parasympathetic branches of the central nervous system. Laboratory evidence and research conducted by Korr show that autonomic function can be modified by a facilitated segment – an articular restriction which is capable of modifying the segmental neuronal firing rate to an extent where changes in function can be experienced locally and at the point of innervation, for example the colon or stomach. Osteopathic manipulative therapy restores normal articular function and breaks the cycle of facilitation, resulting in a stabilisation of neuronal activity and a cessation of pathophysiological reflexes. This is how osteopaths can effectively treat many organic problems, in addition to purely mechanical conditions.[2]

OSTEOPATHIC THEORY AND PRACTICE

However, apart from grasping the general principle of how osteopathy works, it is useful to understand what this principle is based on. There are five important principles which more than adequately capture the essentials of osteopathic theory and practice.

1 The body is an integral unit, a whole. The structure of the body and its functions work together interdependently. Proper balance means health: improper balance can mean susceptibility to disease and illness. Treatment must be directed toward the whole person.
2 The body systems have built-in repair processes which are self-regulating and self-healing in the face of disease. (Note how a simple cut is quickly resolved.) Osteopaths assist the body's healing mechanisms by removing obstacles to its function.
3 The circulatory, lymphatic and nervous systems provide the integrating functions for the rest of the body. Dysfunctions in them frequently lead to congestive disorders, heart disease and a host of stress-related illnesses.
4 The musculoskeletal system does much more than provide framework and support. It is one of the most vulnerable areas responding to stress.
5 While disease may appear in one area of the body, other body

parts may contribute to a restoration or correction of the disease. This is known as referred symptom. Chest pain, for example, need not be due to angina, infection or heart attack; the cause may lie in any number of places, where an osteopath is trained to look.

The key to osteopathic treatment is accurate diagnosis and selective application of therapy. Osteopathic physicians are skilled in both medical and osteopathic diagnosis; thus every patient is thoroughly screened, examined, and orthopaedic and neurological tests are performed, as are systemic valuations when necessary. The two diagnostic approaches are complementary and provide a complete picture of the whole person, in preference to highlighting symptoms.

Once a diagnosis has been reached, it is discussed with the patient with regard to how they might wish to participate in restoring their health. Many patients are very enthusiastic about this opportunity to do something, be it dietary modification, exercise or lifestyle changes. Counselling can often be a powerful and rewarding process at this stage of treatment. For those patients with a more conservative view of their role, the physician will select the appropriate treatment and implement the required changes as needed.

The primary avenues of osteopathic treatments include soft tissue manipulation, muscle energy procedures, high-velocity functional technique and craniosacral therapy. Of these five methods only one involves manipulation, producing the notorious 'pop'. (What should be mentioned here is that osteopathic manipulative therapy does *not* mean that the spine has to be manipulated forcibly and joints made to 'pop'. This misconception has arisen, sadly, from practitioners other than osteopaths cracking their patients' joints and calling it osteopathy.)

High-velocity manipulation is the most widely known and abused method of adjusting vertebrae in the world today. It is practised by doctors, physiotherapists, acupuncturists, naturopaths, masseuses, chiropractors, Indian barbers – in fact anyone confident enough to try it. What good this really does depends entirely on the skill of the practitioner and his reasons for doing it. The technique is excellent for rapid alleviation of a wide range of articular restrictions, but often at a price, including tissue irritation, bruising, injury and, in some unfortunate cases, death.

Anyone considering this method of treatment from someone other than an osteopath or equally skilled practitioner is either very brave or foolish. High-velocity technique should not be regarded as the first choice of osteopaths and it is never used without support. Soft tissue and functional technique are often combined with high-velocity work in order to exclude tissue irritation and maximise functional integration. Along with muscle energy procedures, these three systems allow for a multi-faceted approach to a wide range of disorders, offering a type of treatment which is non-invasive, non-traumatic and, above all, sensitive.

Craniosacral therapy[3] is a particularly sophisticated form of osteopathic care which is continually evolving under the direction of ambitious physicians determined to locate the illusive key to whole-body responses. The ability to do this stems from an appreciation of the existence of the cranial rhythm, an independent physiological entity of 8 to 12 cycles per minute, whose functions are still being explored. It has been observed that the modification of this rhythm in certain patients can have profound effects on systemic functions. In particular, blood pressure has been reduced, respiratory and cardiac rhythms changed, together with gastrointestinal and neuroendocrine function. Research going on in Australia, America and Europe into craniosacral therapy to some extent parallels research into the effects of the use of yogic techniques using biofeedback instruments, for it is in the area of voluntary control over autonomic function that the frontiers of knowledge are expanding.

APPLICATIONS OF OSTEOPATHY

The clinical applications of osteopathy, too, parallel those areas that most wholistic therapies are effective in:

1   Problems of mechanical origin, acute and chronic.
2   Systemic disorders without apparent pathology.
3   Patients who choose not to use medication or surgery.
4   Patients who cannot tolerate medication.
5   Conditions which are not responding to standard medical treatment.
6   Preventative medicine.

OSTEOPATHY AND YOGA

It is, perhaps, in this last area that osteopathy and yoga have most in common, for ultimately osteopathy, like yoga, is concerned with finding a better way of restoring good health and relief from disease. Although their methods may be very different in many ways, functionally they share many common elements which are particularly accessible to yoga practitioners. These include:

1 the functional integrity of the musculoskeletal system;
2 the awareness of articular and soft-tissue restrictions;
3 the relationship between structural abnormality and organic disease.

As osteopaths diagnose and treat their patients via the musculoskeletal system, yoga teachers and practitioners can play a vital role in consolidating and maintaining the health of patients by the selective use of breath and movement. From the osteopath's point of view, however, it would be desirable to cooperate with those teachers who have some training in anatomy, physiology and, above all, biomechanics, as this may facilitate an initial diagnosis of a mechanical problem. The therapeutic use of yoga would require an understanding of the patient's underlying abnormality and which movement would most assist their natural recovery. The long-term benefits of this technique as a self-help exercise cannot be under-estimated.

The relationship between osteopaths and yoga teachers should be based on an appreciation of respective skills. A yoga teacher who has himself/herself been under osteopathic care is more likely to be sensitive to the need for the cautious implementation of their programmes, with particular attention to the fact that the patient feels relaxed and at ease. With effort and cooperation, the relationship between our two disciplines has a positive future. By exchanging ideas, techniques and understanding, there is every reason for osteopaths and teachers of yoga to develop a strong relationship in wholistic health care.

As society approaches the year 2000 the question of health care has become increasingly important, particularly in the light of rising costs and, some would say, decline in services. The range of medical services has, in fact, expanded with new technology and their effectiveness is often questioned, perhaps more so from the point of view of total patient care than simple cure rates. It is interesting to note a general trend developing among patients who ask for two major things from their practitioners – wholistic care and cooperative treatment. Compared with the numbers of patients being quickly classified according to their symptoms and despatched with a prescription, those receiving wholistic care are few and fortunate. But the numbers are growing. The prospect of being referred to someone not practising mainstream medicine may be slim at present, but people are developing a greater awareness of the alternatives available to them. Although the concept of cooperative care – of practitioner and patient forming a team, each assuming a certain degree of responsibility and care for the overall course of treatment – is still very new, even among alternative practitioners, the need certainly exists. If a patient is able to participate actively and cooperate in the treatment, this will often make the difference between success and failure and will frequently accelerate recovery times significantly.

Health care as it is understood and practised now in our major democracies must change. It must evolve into a more open and integrated system which can serve the total individual in a manner beneficial to both patients and practitioners of all disciplines, for only a truly wholistic approach can ensure our healthy survival into the twenty-first century.

REFERENCES

1   I. Korr et al., The Physiological Basis of Osteopathic Medicine, Postgraduate Institute of Osteopathic Medicine and Surgery, New York, 1970.
2   B. Jones, The Difference a DO Makes, Times Journal, Oklahoma, 1978.
3   J. Upledger, Craniosacral Therapy, Eastland, Chicago, 1983.

16 *Yoga as Psychotherapy*

MADAN N. PALSANE

In view of the growing trend towards complementary healing methods, it seems appropriate to discuss psychotherapy, or treatment of the personality, since yoga has played an important therapeutic role in bringing Eastern and Western schools of thought closer together.

Several definitions of psychotherapy have been put forward by various people working the field of human behaviour, but most of them seem to agree that it is a planned intervention by trained specialists to help people with behavioural and emotional problems of psychogenic origin to overcome them and restore normal functioning. However, the specialists concerned put different emphases on aspects of growth or improvement, depending on their theoretical positions.

But whatever the approach of the particular therapist, the main features of psychotherapy are:

1   a person called the client, with an emotional or behavioural problem;
2   a trained expert and integrated person, called the psychotherapist;
3   the expert–client interaction and relationship;
4   deliberate effort on the part of both to restore normality;
5   the techniques used based on psychological facts and theories.

How the psychotherapy is administered also varies. It may be in complete privacy or in groups, the latter being favoured by some therapists for its facilitative aspects. Some therapies analyse the individual's behaviour at the conscious levels, while others give much importance to the unconscious, and work at the in-depth levels. Some psychotherapists believe in taking a large responsibility in directing the course of change in the client's behaviour, while others are non-directive. Some place emphasis on the past and future of the client; others deal with the present and emphasise the 'here and now' philosophy and lifestyle.

Individual counselling, clinical interviews, group dynamics, encounter groups, psychodrama and even meditation are among the many procedures used. Most psychotherapists use more than one of these approaches or an integration of several.

Although it is difficult to talk about goals of psychotherapy in view of the fact that each psychotherapist aims towards a different objective according to his interpretation and evaluation of each case, there are some common features to the process. Psychotherapy is intended to help the individual:

1 gain relief from anxiety;
2 establish feelings of adequacy of self;
3 acquire self-integration and personal maturity;
4 develop effectiveness in life;
5 improve interpersonal relationships, including the ability to give and receive love;
6 increase adjustment to society and culture;
7 become increasingly self-sufficient and minimise dependence on the physical and social environment;
8 increase awareness and act decisively on the basis of awareness;
9 learn self-acceptance and develop adequate self-worth and self-regard;
10 develop meaning in life around a central value and an integrated system of rational beliefs.

All these goals are intended to help the individual grow, become effective and gain control over his or her life. This is known as normality, adjustment, maturity or reintegration and implies that anyone approaching a psychotherapist is a sick person trying to get back to normal functioning.

This notion of normality varies according to the school of thought. They attribute abnormality to different causes and describe it in different terms, but for convenience they can be classified into three groups:

1   the medical model;
2   the statistical model;
3   the utopian model.

The medical model

This is also called the disease model, and has been borrowed from the field of physical health. The history of modern medicine is firmly rooted in this concept. According to this viewpoint, health is the absence of disease (dis-ease) or the absence of any pathological symptom. This is a negative approach, particularly visible in the thinking of Freud and the behaviourists. According to Freud, unfulfilled and repressed desires, conflicts between the demands of the id and the superego and the preponderance of various ego-defence mechanisms give rise to pathological symptoms culminating in mental disorders.

The behaviourist, on the other hand, concentrates on the superstratum of behaviour. He believes that mental disorder is the outcome of learning and development of faulty behaviour patterns which eventually lead to pathological symptoms and mental disorders.

The goal of psychotherapy, according to this model, is to free the individual from deviant forms of behaviour and thus restore the individual to normal functioning, but this is still a negative approach.

The statistical model

This includes the conceptualisations of Jung, Allport, Cattell and some social learning theorists. Some, like Jung, follow the type or trait approach to personality; that is, normality consists of a balance between two extremes. Allport, Cattell and Eysenck also described personality in a similar way, but their concept of normality was arrived at through statistical averages for specific population groups. Social learning theorists considered that the norm is based on the expectations of the majority population and in imitation of set performance standards.

The utopian model

This model is hard for psychologists rooted in positivistic theory to conceptualise within the present framework of Western science, but both Abraham Maslow and Carl Rogers made the attempt. Their respective visions of a healthy personality were neither restricted by the notion of disease nor by the statistical average human performance. They were perhaps idealists.

Maslow believed that most individuals end up achieving short of their potential. He urged people to venture into greatness, to rise from the mere satisfying of lower-level needs like hunger and sex, safety and belongingness, and to reach up to more obviously human levels like self-esteem and self-actualisation.

Rogers conceived of a fully-functioning person in more or less the same terms. A person should try to become himself by advancing towards autonomy, freedom and identity. And, as this process of becoming oneself is supposed to be enjoyable, this is considered happiness and self-fulfilment.

THE 'HEALTHY' PERSONALITY

With few exceptions, then, Western psychotherapists aim primarily at symptom reduction and the restoration of the 'normal' personality. Imprecise as the latter definition is, we can see from the following list of characteristics that the 'healthy' personality must be able to relate to oneself and to others.

Attitudes towards oneself

The normal, healthy personality aims to:

1 acquire and maintain physical fitness by following proper diet, rest, exercise, cleanliness and hygiene;
2 know how to avoid strain and to relax the body and mind at will;
3 grow through participation rather than being a passive spectator;
4 strive to reach higher and yet higher levels of satisfaction and attainment;
5 enjoy living in all respects, especially the present, and to feel that life is meaningful;
6 develop various interests and hobbies or useful pastimes;
7 acquire realistic knowledge of the world, decision-making skills and wisdom to pass judgments;

8 develop realism in goal-setting and the ability to plan how to attain set goals;

9 develop an objective perception of self and one's capacities and maintain a consistency between goals and capacities;

10 develop adequate self-esteem and self-confidence;

11 have emotional stability and maturity, control impulses and tendencies towards over-sensitivity, nervousness, anxiety and insecurity; be optimistic and learn not to dwell on unpleasantnesses; tolerate frustration and quell feelings of inferiority or frustrated strong urges;

12 have a balanced personality, inner harmony, integration, autonomy, rationality, a sense of humour, naturalness and an ability to meet crises;

13 develop a personal philosophy, and look on adversities philosophically and with patience;

14 enjoy work, have a repertory of useful skills and keep busy.

Attitudes towards others

A normal, healthy person can:

1 share feelings of love and belongingness;

2 develop understanding of others, maintain flexibility in inter-personal interactions with almost anyone;

3 develop some intimate and lasting friendships;

4 develop trust in others and a healthy attitude towards the world at large;

5 develop tact and tolerance, avoiding arguments and self-ishness;

6 look beyond the self to society and culture, but not be bound by them.

TRANSCENDENCE AND YOGA

Eastern thought is not so well codified, but can roughly be represented by the ideas incorporated in Hinduism, Zen Buddhism and Sufism. The common orientation of these schools of thought is *transcendence*. The individual is supposed to develop toward freedom, autonomy, self-realisation *and* super-conscious levels. There is a rigorous ethical code to be followed so that one can refine oneself sufficiently to attain the goal. In this sense, then, religion could be described as an applied psychology enterprise in self-development.

Yoga advocates a similar path to transcendence but does not presuppose religious belief. Development is seen as progress toward something *beyond* oneself, so that integration is not only a goal within oneself but extends to unity with a transcendental self or totality. Yoga could be described as wholistic, as opposed to analytical, but it is also individual and phenomenological.

According to yoga philosophy, an individual evolves out of nature, phylogenetically as well as ontogenetically. He is, therefore, a part of the system of nature. And this is what the concept of normality has developed from. Humans are considered abnormal when they violate the laws of nature. Yoga evolved as a moral science or code of ethics to restrain man from using his freedom unwisely and thereby destroying nature and, in the process, his own chance of happiness.

The yogic path is very gradual and each skill or practice is acquired through tiny steps like those in programmed learning. Regular and sustained effort creates stability of thought, feeling and action. Ego involvements, and the resulting emotions of attachment and aversion, thought to be the source of all our misery, are modified, making it possible for the individual to let go of certain biases and to be more open to receive information, people and experiences with an expanded consciousness and perception of the world. As the individual's lifestyle becomes more in tune with the yogic ideology, so he becomes more rational and consistent in his behaviour. A relaxed mind is more conducive to creativity and to achieving realistic goals without undue stress. A balanced outlook allows the individual to enjoy the process of achieving without being too attached to the outcome. To enjoy the present means he enjoys himself, the person that he is. This is in itself a psychologically healthy orientation, shared by the humanists and existentialists of Western psychotherapy.

While learning the typical yogic skills and attitudes, step by step and gradually, there are other very important and psychologically significant outcomes such as self-confidence and self-esteem, acquired as a result of experiencing success at each step. These are not merely byproducts of the process; they are highly desirable goals in their own right and it is possible that these are of greater importance than the more apparent tangible skills and behaviours.

The self

But yoga is no more the sum total of its steps than the individual is the sum total of his characteristics. The whole system has a character of its own that differs greatly from any other, particularly any Western ideology. Its essential feature is that it is a system of *self-study*. The understanding of the self, and the self-related science from which the practices have developed, grows to the extent that the student evolves in the methodology of its study. Intuition and insight play a larger role than objective verification and validation. In simple terms, it may be said that the subject matter of Western psychology is behaviour seen *analytically* in terms of its elements and their interrelationships, while the subject matter of Indian psychology, as seen in yoga, is the *self as a whole* as it evolves to reach the fully-realised state. As the subject matter of yoga is experiental in terms of one's development, character and relationship with the external world, there is greater emphasis on methodology than on the generalisation and theory of the subject matter. It is in this sense, then, that the yogic approach is highly individual and begins with the study of self, without dealing with any particular element of behaviour or thought.

The term 'self' in yogic or Indian traditional thought also differs from the 'self' of Western psychology. The self of yoga is not a personal identity, but an evolved state of eternal bliss, beyond any personal attachments, aversions, ego involvements or false perceptions, known as *samadhi* or *nirvana*.

Conscious training through concentration and meditation teaches the individual how to control his thoughts and voluntarily eliminate unwanted ones, just as movements and muscle tensions can be eliminated by relaxation procedures. This reduces the causes of arousal and the concomitant bodily disturbances that accompany certain thoughts. The practice of breathing and physical postures (asanas) provides gradual control over the involuntary autonomic functions which are basic to arousal, while various other practices, including bandhas and kriyas which elicit reactions at certain threshold points, help to raise threshold levels to the point where sufficient control over these involuntary reactions can be experienced. Rapid breathing, sweating, tremors, blood circulation changes, coughing and sneezing are among the kind of reactions that these bandhas and kriyas elicit which closely resemble the symptoms of some physical and mental disorders. In this way yoga helps to immunise against disease by enabling one to tolerate symptoms and irritations and so giving a sense of mastery over the disease process, thus inhibiting its progress.

Stress and suffering

Many elements of yoga philosophy and practice are common to various religions and beliefs. All these systems possibly have their origin in the fact of suffering. Without suffering, ethics perhaps would not have evolved. The search for God may also have originated in the effort to deal with stress and suffering. Most religious sects and leaders have their strongest following among the sick and distressed and there is some evidence that the comfort derived from religious belief helps to alleviate stress.

Stress is the rampant malady of our age, causing a variety of physical and mental disturbances. Stress-related disorders seem to respond best to relaxation procedures and yoga is an effective remedy in this instance. This may be because, just as stress is a non-specific response to threat, yoga is a non-specific remedy working at a wholistic level.

The value of yoga as a means of adaptability or improvement of mental health can be examined from a more recent perspective. According to Taylor's cognitive theory of adaptation, adaptability depends on one's capacity to create and sustain illusions. Not only does this allow the individual to attach personal meaning to the situation to be adapted to, but it also gives a sense of control and enhances self-esteem as a result of the experience. Fasting is an interesting instance. It is essentially deprivation, but people undergo a fast, especially if they are motivated by religion, without complaint. All three aspects of adaptation are present in this example: firstly, fasting is viewed as meaningful in a certain religious and/or value context; secondly, increasing the fasting period, step by step, gives a sense of mastery over the situation; and thirdly, it enhances self-esteem, quite apart from any beneficial effect on health and well-being.

The discipline of yoga as a whole can be viewed in much the

same way. Although it appears to deprive the student of freedom, it has a positive value based on a system of rationalisations that allows the individual to cope with whatever rigours it presents and without any negative side-effects. The utility of yoga as a therapy therefore is self-evident for, in coping with situations that arise in yoga practice, the student is better prepared to deal with the world at large.

CONCLUSION

Within the confines of limited space, this is but a sketchy outline. Not only is it difficult to draw a comparison between Eastern and Western approaches because they are based on different concepts, but yoga is also hard to study because it is so subjective. It is probably for this reason that there has been greater interest and research into the study of tangible overt behaviour, especially with animals, which are much less complex than human beings. However it is time that scientists took up the challenge, for there is much to be learned to our advantage from our traditions of ancient wisdom, which the modern world, with all its technological knowledge, has not truly fathomed.

PART IV

The Potential

17 *The Integration of Yoga Therapy*

M. L. GHAROTE and
MAUREEN LOCKHART

People have different attitudes towards illness. Some people 'ignore it' in the hope that it will 'go away'; others make a career out of it. Most of us struggle valiantly to live our lives in spite of it and, if we can, try to put the knowledge we gain from the experience to good use in helping others. Yoga helps us to see illness in its proper perspective. It rids us of the fear of 'falling apart', of becoming helpless hopeless invalids. It makes us realise that illness is an experience like any other, a part of our existence that we cannot ignore, and we have to learn to integrate it so that we can grow beyond it. Yoga helps us to learn the fundamental principles of life, drawn from the inner experiences of countless thousands who have learned to integrate them into their daily lives. The skills that we learn through yoga help us to change the self-destructive habits that prevent us from becoming vibrant, happy human beings. It releases us from self-imprisoning ideas that limit our capacities and frees us to expand and enjoy our abilities and relationships.

It would be unrealistic to say that yoga therapy is a cure-all for every kind of disease, but it is amazing how many illnesses *do* respond to yoga. Yoga makes an incredible number of people throughout the world feel good every day of their lives. When they compare how they used to feel before they discovered yoga, they realise that they were not really living, merely existing. It makes people confident that they are capable of running their own lives in their own way. The more people there are who develop this level of maturity, the less of a burden they are on their family and society. When people are motivated by integrity rather than selfishness, their vision is expanded. They encompass more of the world in their outlook and learn to care about those less fortunate than themselves. They feel responsible for the world *as a whole*.

People with integrity are badly needed in the world today. They are needed in every walk of life. They are needed to inspire others to develop their own potential and stand on their own two feet. They are especially needed in the healing professions, for the care of the sick requires a special kind of awareness that comes from integrity. Yoga therapists are a valuable asset in the healing arts for it is axiomatic that they develop integrity. Integration is the *purpose* of yoga.

There is a great deal to be done in developing the therapeutic value of yoga into an effective healing art. Standards have to be set for the proper training of yoga therapists; appropriate methods of application have to be developed. And the integration of yoga therapy into a wholistic health system that truly cares for the sick person is a pressing necessity.

In the last few years, we have made a beginning. Yoga awareness has spread so rapidly that there are few places in the world where it is unheard of. The need to conserve our diminishing outer resources necessitates the integration of yoga into our lives to help us develop our inner resources. We need to learn to desire less and share what we have if we are all to survive with dignity.

We do have the resources. We even have the methods to develop these resources. What we have to do now is to put them into practice. The art of survival is yoga. Yoga is the Art of Survival.

Appendix 1 *The Editors and Contributors*

THE EDITORS

M. L. Gharote

M. L. Gharote is the assistant director of research and the principal of India's Kaivalyadhama College of Yoga and Cultural Synthesis, which houses the world's oldest established scientific laboratory conducting experiments in yoga research and one of the first established yoga hospitals. He is a doctor of philosophy, with degrees in education; a Sanskrit scholar; the author of several books on the many aspects of yoga practice; and has contributed to many leading yoga journals and magazines. He is well known to yoga students and teachers in several countries, including Italy, Germany, Britain and South America, through his popular lecture tours and demonstrations. He has been instrumental in stimulating interest in yoga therapy and ayurveda in several countries and has been invited to prepare the curriculi for university courses, as well as promoting seminars and conferences on research into yoga and preventive medicine.

Maureen Lockhart

Maureen Lockhart first discovered yoga in the early 1960s at the age of 16 in her native Scotland. It not only cured her of chronic bronchitis, which she'd had since childhood, but later helped her recover from a severe spinal injury. While working as a journalist in London, she took her British and international yoga teacher's diplomas, became a staff teacher for the ILEA, the first yoga teacher on the faculty of the London Mime Centre and for many years taught people of all ages and professions, including doctors and physiotherapists. Her yoga experience prompted her to train in biofeedback with C. Maxwell Cade, in postural integration, humanistic psychology and naturopathy and to specialise as a freelance editor in wholistic therapies for several leading London publishers and to contribute to journals and magazines. Since 1979 she has been living in the hills of western India where she continues to write, teach yoga to people of all nationalities and has a private practice in Bombay.

THE CONTRIBUTORS

Lilla Bek

Lilla Bek is well known throughout Britain for her healing gift which she discovered through yoga. Since the early 1970s she has been helping people to develop their own awareness in her yoga classes and through her psychic readings, lectures, tapes and published books. Through her ability to see the energies around people, which indicate their inner state of physical, mental and spiritual well-being, she is able to help healers and their patients tap their own resources of healing and to understand their present life and needs. She is keenly aware of the need to explore these gifts more fully and collaborates with scientific researchers to discover their meaning. She is also a consultant to the *Yoga Times*.

M. V. Bhole

M. V. Bhole is a doctor of medicine and joint director of research at Kaivalyadhama Yoga Research Institute, Lonavla, India. Although he is interested in all aspects of yoga practice, he is known to many teachers and students in several countries through his lectures on pranayama and its effect on psychosomatic and respiratory disorders, on which he has written several papers.

Mark Blows

Mark Blows is a clinical psychologist in private practice in Sydney, Australia. He has worked in several psychiatric centres, especially on family therapy. He is particularly interested in the effects of stress on the family group in large competitive cities and has been using meditation as a tool in therapeutic practice. His investigations have taken him to several countries, including India, to look into the effects of yoga on the nervous system and he is currently preparing a book on his own eclectic system of relaxation therapy.

Barbara Brosnan

Barbara Brosnan graduated in modern languages at London University before becoming a nurse at the Middlesex Hospital. She went on to obtain her MB and ChB from Bristol University and worked for 13 years on the staff of the Bristol Royal Infirmary. In 1962 she started the Servite House in Ealing, a home for the physically handicapped, which she still runs. Under her guidance, the home is moving steadily towards the yogic concept of life, and the practice of hatha yoga – with modifications – is widely fostered. Workshops for teachers wanting to teach yoga to those with a physical handicap are held regularly at Servite House, while the residents, together with Doctor Barbara, give demonstrations all over the country. She has recently published a practical guide which gives valuable insight into working with the handicapped through yoga.

Ctibor Dostalek

Ctibor Dostalek is director of the Academy of Sciences in Prague and a leading Czech scientist. He is particularly interested in the physiological aspects of yogic techniques, including postures, breathing and meditation. His research has taken him to several countries, including India, where he has made extensive studies. He has contributed to many scientific journals and lectured at several symposia on his experimental work.

Stanley Jacobs

Stanley Jacobs was born and educated in Scotland. He did his medical training in Glasgow and London and has worked in Norway and Israel as well as the UK. He was responsible for introducing wholistic methods of treatment, including yoga, to the University College Hospital, London, psychiatric inpatient department and to the neurosis unit at Shenley Mental Hospital, St Albans, in the early 1970s. He lectures regularly to professionals and to the public on the principles and philosophy of ayurveda. He is presently working as a consultant psychiatrist, psychotherapist and counsellor for several London boroughs and the ILEA. He is an executive member of the British Holistic Medical Association and a founder member of the Society for Ayurveda.

Roman Maslak

Roman Maslak is the director of clinical studies for the International Colleges of Osteopathy in New South Wales, Australia. He also lectures extensively to several osteopathic guilds and societies as well as conducting his own private practice and research into specialised areas, such as 'the functional dynamics of the cervico-thoracic junction in relation to upper limb dysfunction'. He believes the future of therapy lies in a wholistic system in which osteopathy, yoga and many other therapies work together to provide the best possible form of patient care.

Pedro de Vicente Monjo

Pedro de Vicente Monjo is the director of the Institute of Relaxation Technique and Yogic Therapy at the University of Seville's Faculty of Physical Medicine, Spain. He is a leading cardiologist and winner of two gold medals from the International Congress of Scientific and Educational Films for his film on *Yoga and Respiration* in 1978 and 1982. He is a trained yoga teacher from Kaivalyadhama, Lonavla, an international teacher of sophrology, a training doctor for the Spanish Federation of Scuba Divers using yogic techniques, and an international delegate from the Spanish government in problems with drug trafficking and drug addiction. He has done considerable research into the therapeutic use of yogic techniques.

Hiroshi Motoyama

Hiroshi Motoyama is director of the Institute for Religious Psychology, Tokyo, Japan. He has been investigating the effects of various complementary fields of medicine but has a special interest in the electromagnetic field of the body and the parallels between the chakras and nadis of yoga and the points and meridians of acupuncture. His contributions to this field of research are well known through his several publications and his venerable positions as president of the International Association for Religion and Parapsychology and 'accademio ordinario' of the Academia Tiberina, Rome. Apart from being a PhD in both philosophy and physiological psychology and a scientist, Dr Motoyama is also a shinto priest, an acupuncture specialist and a yoga teacher.

Madan N. Palsane

Madan N. Palsane is professor of psychology at the Department of Psychology, University of Poona, India. Before becoming head of the department of psychology at Poona, he worked as a research fellow, a student counsellor and lectured in many aspects of psychology in several other Indian universities and at seminars, symposia, workshops and summer camps. In addition to his 20 published research papers and some 35 other papers, monographs and books, he has contributed *The First Survey of Research In Education*, a major reference work. His present research interest is focused on the problems of stress and related phenomena, and particularly in the integration of a modern perspective with traditional Indian thought.

Peter Valance Sandhu

Peter Valance Sandhu was born in Bombay and lived in England from the age of eight, where he was educated in Devon and London. He trained in psychotherapy and took his doctorate in hypnotherapy and was in private practice in London for several years. The major part of this work was psychoprophylactic treatment using yoga-derived methods for painless childbirth. Since returning to India in 1970, he has been actively involved in social development in rural areas, particularly in developing self-sufficiency and self-awareness among the rural poor.

Jane Simons

Jane Simons was born in Zimbabwe and spent her school years there. Having Czechoslovakian parents and living in Africa heightened her awareness of the cultural differences between people and the wide spectrum of values that separated them. She was particularly struck by the different attitudes towards childbirth in different societies. During her training at Guy's Hospital, London, she developed an interest in obstetric physiotherapy, which was intensified by her work in England and later in Australia, where she is now settled. Her own birthing experiences in a new country made her acutely aware of the sense of separation many young mothers feel during pregnancy and early motherhood. Her work as a staff physiotherapist at a New South Wales physiotherapy centre for maternal and child health enables her to help mothers-to-be to prepare for birth through confident self-management.

Velta Snikere Wilson

Velta Snikere Wilson was born in Latvia and studied philosophy at the University of Riga, where her father was professor of medicine. After serving as an interpreter for the British Army in Austria from 1945 to 1946, she trained and became a member of the Chartered Society of Physiotherapists in London in 1954. As a child, she spontaneously practised physical and mental yoga and later trained with several well-known teachers including Sir Paul Dukes and Ram Gopal, with whom she also trained in Indian dancing, performing as a member of his troup at several European festivals. She has been teaching yoga since 1965 and has run British Wheel of Yoga teacher training courses since 1972. She studied autogenic training with Dr Malcolm Carruthers and now teaches it in her private practice. As well as her many articles on yoga which have appeared in various journals, she is also acclaimed as one of the foremost contemporary Latvian poets.

Anna Wise

Anna Wise is co-founder and director of the Evolving Institute in Boulder, Colorado, USA, which she created in 1982 with her husband, dance teacher and now practising acupuncturist, Jym MacRitchie, to integrate and utilise the diverse but complementary aspects of their work in evolution. In 1971 she left America to travel, study and work in Europe. During the 10 years she lived in England, she trained in biofeedback and consciousness development with C. Maxwell Cade and became a director of the Institute for Psychobiological Research in London. As co-founder of the Natural Dance Workshop, she led courses, workshops and training programmes throughout Europe and the United States. Her MA is in humanistic psychology and she was a founder board member of the European Association for Humanistic Psychology. Apart from teaching the Awakened Mind Programme and meditation classes she also has a private practice in Boulder.

Appendix 2   *Bibliography*

GENERAL

Books

Swami Ajaya, *Meditational Therapy*, Himalayan International Institute of Yoga Science and Philosophy, Chicago, 1977.

Swami Rama, Ballentyne and Swami Ajaya, *Yoga and Psychotherapy*, Himalayan International Institute of Yoga Science and Philosophy, Chicago, 1976.

Swami Rama Ballentyne, Phillip Nuernberger, Charles Bates and Jagdish Dua, *Therapeutic Value of Yoga*, Himalayan International Institute of Yoga Science and Philosophy, Chicago, 1979.

Lilla Bek, *What Colour Are You?*, Turnstone, Wellingborough, Northants, 1981.

Herbert Benson, *The Relaxation Response*, Avon, New York, 1976.

Steven F. Brena, *Yoga and Medicine*, Julian, New York, 1972; Penguin, London, 1973.

Barbara Brosnan, *Yoga for Handicapped People*, Souvenir, London, 1982.

Brian Edward Bunk, *Effects of Hathayoga and Mantra Meditation on the Psychological Health and Behaviour of Incarcerated Males*, PhD thesis, University of Texas Health Science Centre, Dallas, 1978.

Ramiro A. Calle, *Principios de Yogoterapia – Yoga Ciencia de La Salud*, Piramide, Madrid, 1979.

K. K. Datey, M. L. Gharote and Soli Pavri, *Yoga and Your Heart*, Jaico, Bombay, 1983.

R. K. Garde, *Principles and Practice of Yoga Therapy*, D. B. Taraporewala, Bombay, 1973.

Swami Gitananda, *The Correction of Breathing Difficulties by Yoga Pranayama*, Satya, Pondicherry, 1972.

Hans Jacobs, *Western Psychotherapy and Hindu Sadhana*, Allen & Unwin, London, 1961.

O. P. Jaggi, *Yogic and Tantric Medicine*, 2nd edition, Atmaram & Sons, Delhi, 1979.

Stephen Jeffrey Johnson, *Effects of Yoga Therapy on Conflict Resolution, Self-Concept and Emotional Adjustment*, PhD thesis, University of Southern California, 1974.

Swami Kuvalayananda and S. L. Vinekar, *Yogic Therapy – Its Basic Principles and Methods*, Central Health Education Bureau, Ministry of Health, New Delhi, 1963.

Maureen Lockhart, *Yoga for Youth: A Young People's Guide To A Healthy Life*, India Book House, Bombay, 1986.

John Merer, *Asthma and Yoga*, Bihar School of Yoga, Monghyr, 1975.

G. S. Mukerji and W. Spiegelhoff, *Yoga und unzere Medizin – Arztliche Anleitungen zur Yoga ubungen*, Hippokrates, Stuttgart, 1963.

Barte Nhi, *Yoga et Psychiatrie*, Tete de Fuilles, Paris, 1972.

Kumar Pal, *Yoga and Psychoanalysis*, Kumar, New Delhi, 1966.

Proceedings of the Seminar on Yoga, Science and Man, Central Council of Research in Indian Medicine and Homeopathy, New Delhi, 1976.

B. V. Sathaye, *Vihar Chikitsa Through Yoga Procedures – Rationale, Scope and Limits*, PhD thesis, Poona University, 1984.

Bill Schul, *The Psychic Frontiers of Medicine*, Coronet, London, 1977, especially Chapter 9, 'The healing reality of meditation'.

J. Sedivy, *Yoga as Seen by a Physician* (Czech), Olomouc, Prague, 1979.

Swami Shankaradevanand, *The Effects of Yoga on Hypertension*, Bihar School of Yoga, Monghyr, 1978.

Swami Shankaradevanand, *Yoga Management of Asthma, Blood Pressure, Diabetes*, 2nd edition, Bihar School of Yoga, Monghyr, 1979.

S. L. Sharma, *Yoga Technique of Psychotherapy*, Metropolitan Book Co., New Delhi, 1979.

K. N. Udupa, *Disorders of Stress and Their Management with Yoga*, Benares Hindu University Press, Varanasi, 1978.

Jurg Wunderli, *Yoga und Medizin – Ein Arzt uber den Geistigen Yoga und seine Beziehungen zur Heilkunde*, ABC Verlag, Zurich, 1964.

M. Yamazaki, *Yoga and Psychology (Yoga as a Living Therapy)*, Fukui University Press, 1977.

Articles

S. Ao, T. Noda and J. R. M. Goyeche, 'The role of yoga therapy on a psychosomatic ward', paper read at IV Cong. Int. Coll. Psychosom. Med., Kyoto, 1977.

A. Arbe, 'That new tranquilliser called yoga', *Med. Times*, 100, Sept., 1972, p. 107 *et seq.*

H. Armstrong, 'Yoga: one physician's experience', *Can. Med. Ass. J.*, 118, 1978, p. 992.

B. Auriol, 'Group yoga therapy. Indications and contraindications for a new psychosomatic technique' (French), *Psychother. Psychosom.*, 20, 1972, pp. 162–8.

H. Benson *et al.*, 'Relaxation response – bridge between psychiatry and medicine', *Med. Clin. N. Am.*, 61, 4, 1977, pp. 929–38.

J. Bourdouxhe, 'The practice of yoga in the prevention and treatment of occupational diseases of the dental practitioners' (French), *J. Rev. Belg. Med. Dent.*, 24, 1969, pp. 23–48.

O. P. Bhatnagar, 'The effect of yoga training on neuromuscular excitability and muscular relaxation', *Neurol. India*, 25, 4, 1977, pp. 230–2.

S. N. Bhavasar, 'Pathogenesis with special reference to yoga', Souvenir of Yogis' Conference of Yococen (Yoga Co-ordination Centre), New Delhi, 1981, pp. 79–82.

M. V. Bhole and P. V. Karambelkar, 'Yoga practices in relation to therapeutics', *Yoga Mimamsa*, 14, 3 and 4, 1971–2, pp. 27–34.

B. Boelsen, 'Yoga, a healing power within us', *Krankenpflege*, 29, 12, 1975, pp. 474–83.

P. Carrington and H. S. Ephren, 'Clinical use of meditation', *Current Psychiatric Therapies*, 15, 1975, pp. 101–8.

V. A. Dhurandhar, 'The incurable and yoga', *Journal of the Yoga Institute*, 14, 6, 1969, pp. 84–7.

G. Ferrari and P. Roberti, 'Yoga e yogaterapia; verso un nuovo approccio psicosomatico', *Intervento Preordinato al XXXV Congresso della Societa Italiana di Psichiatria*, Cagliari, 1982.

A. P. French and Joe P. Tupin, 'Therapeutic application of a simple relaxation method', *American Journal of Psychotherapy*, 28, 1974, pp. 282–7.

J. R. M. Goyeche, 'Yoga: clinical observations and somatopsychic principles', paper read at IV Cong. Int. Coll. Psychosom. Med., Kyoto, 1977.

J. R. M. Goyeche, 'Yoga as therapy in psychosomatic medicine', *Psychother. Psychosom.*, 31, 1–4, 1979, pp. 373–81.

J. R. M. Goyeche, Y. Ago and Y. Ikemi, 'Breathing and psychosomatic medicine', *Asian Med. J.*, 21, 1978, p. 674.

J. R. M. Goyeche and Y. Ikemi, 'Yoga as potential psychosomatic therapy', *Asian Med. J.*, 20, 1977, p. 90.

J. Henrotte, 'Yoga et medicine', *Medica*, 69, 1968, p. 2.

J. Henrotte, 'Yoga and biology' (French), *Atomes*, 24, 1969, pp. 283–92.

J. Hoenig, 'Medical research on yoga', *Confinia Psychiatrica*, 11, 1968, pp. 69–89.

H. J. Jhalla, 'Some viewpoints on medical benefits of yoga', *J. of Yoga Institute*, 17, 1972, p. 115.

L. N. Kurulkar and J. Mehta, 'Yoga – the key to psychosomatics', paper read at IV Cong. Int. Coll. Psychosom. Med., Kyoto, 1977.

M. G. Mokashi, 'Yoga and physiotherapy', paper read at VII International Congress of the World Federation for Physical Therapy, Montreal, 1974.

W. P. Morgan, 'Use of exercise as a relaxation technique', *J. Sc. Med. Assoc.*, 75, Nov., 4, 1979, pp. 596–601.

Hiroshi Motoyama, 'Yoga and oriental medicine', *International Association of Religious Psychology*, Research for Religion and Parapsychology Journal, 5, 1979, p.1.

T. Noda, S. Ao and J. R. M. Goyeche, 'Yoga therapy – a summary of case studies', paper read at IV Cong. Int. Coll. Psychosom. Med., Kyoto, 1977.

K. Porkodi, S. Subramanian and T. S. Kanaka, 'Effects of yoga and meditation in health and disease', in *Proc. First National Conference on Yoga, Science and Society*, Benares Hindu University Press, Varanasi, 1979, pp. 104–6.

Drocanegra R. Prerez, 'Yoga et physiotherapie' (Spanish), *Medicine Tropicale*, 38, 1962, pp. 273–82.

M. Shaffi, 'Adaptive and therapeutic aspects of meditation', *Int. J. Psychoanal. Psychotherapy*, 2, 3, 1973, p. 364.

Sarada Subrahmanyam, 'The science of yoga therapy', Souvenir of Yogis' Conference of YOCOCENE, New Delhi, 1981, pp. 37–9.

Sarada Subrahmanyam, 'Yoga and psychosomatic illnesses', *The Yoga Review*, 2, 3, 1982, pp. 149–59.

N. S. Vahia, D. R. Doongaji, D. V. Jesta, S. Ravindranath, S. N. Kapoor and I. Ardhapurkar, 'Psychophysiologic therapy based on the concepts of Patanjali', *Am. J. Psychotherapy*, 27, 1973, pp. 557–65.

SPECIFIC DISORDERS

Asthma and other respiratory disorders

A. B. Alexander, 'The immediate effects of systematic relaxation training on peak expiratory flow rates in asthmatic children', *Psychosom. Med.*, 34, 1972, pp. 388–94.

A. B. Alexander *et al.*, 'Effects of relaxation training on pulmonary mechanics in children with asthma', *J. Appl. Behav. Anal.*, 12, 1, Spring, 1979, pp. 27–35.

S. Anandananda and N. Varandani, 'Therapeutic effects of yoga in bronchial asthma', in *Proc. Seminar on Yoga, Science and Man*, Central Council of Research in Indian Medicine and Homeopathy, New Delhi, 1975, pp. 157–71.

M. V. Bhole, 'Treatment of bronchial asthma by yogic methods – a report', *Yoga Mimamsa*, 9, 3, 1967, pp. 33–41.

M. V. Bhole, 'Yogic treatment of chronic rhinitis and sinusitis', *Maharashtra Medical Journal*, 17, 1970, p. 8.

M. V. Bhole, 'Rationale of treatment and rehabilitation of asthmatics', in S. Digambarji (ed.), *Collected Papers on Yoga*, Kaivalyadhama, Lonavla, 1975, pp. 105–14.

M. V. Bhole and M. L. Gharote, 'Effect of yogic treatment on breath holding time in asthmatics', *J. Res. Ind. Med. Yoga and Homeo.*, 13, 2, 1978, pp. 1–4.

M. V. Bhole and P. V. Karambelkar, 'Effect of yogic treatment on blood picture in asthma patients', *Yoga Mimamsa*, 14, 1 and 2, 1971, pp. 1–6.

J. R. M. Goyeche *et al.*, 'Asthma: the yoga perspective. The somatopsychic imbalance in asthma: towards a holistic therapy', *J. Asthma Res.*, 17, 1980, p. 111.

G. B. Gupta, G. C. Sepaha, I. Menon and S. K. Tiwari, 'Effects of yoga on bronchial asthma', *Yoga*, 17, 2, 1979, pp. 29–33.

Ron Honsberger and Archie F. Wilson, 'Transcendental meditation in respiratory kinesio-therapy in children' (Rumanian), *Pediatria*, 22, 1973, pp. 155–60.

P. B. Rajput, 'Rhinitis and yoga', *Journal of the Yoga Institute*, 17, 2, 1971, pp. 25–7.

A. V. N. Appa Rao, D. R. Krishna, T. V. Ramanakar and M. C. Prabhakar, 'Jala neti – a yoga technique for nasal comfort and hygiene in leprosy patients', *Leprosy in India*, 54, 4, 1982, pp. 691–4.

M. K. Tandon, 'Adjunct treatment with yoga in chronic severe airways obstruction', *Thorax*, 33, 4, August, 1978, pp. 514–17.

A. F. Wilson, R. Honsberger, J. T. Chiu *et al.*, 'Transcendental meditation and asthma', *Respiration*, 32, 1975, pp. 74–80.

Cancer

J. S. Bolen, 'Meditation and psychotherapy in the treatment of cancer', *Psychic*, 4, 6, 1973, pp. 19–22.

Ainslie Meares, 'Regression of cancer after intensive meditation', *Med. J. Austr.*, 2, 1976, p. 184.

Ainslie Meares, 'Regression of osteogenic serrome metastases with intensive meditation', *Med. J. Austr.*, 2, 9, 1978, p. 433.

Ainslie Meares, 'Remission of massive metastasis from undifferentiated carcinoma of the lung associated with intensive meditation', *J. Am. Soc. Psychosom. Dent. Med.*, 27, 2, 1980, pp. 40–1.

S. Secheny, 'Regression of cancer of the rectum after intensive meditation' (letter), *Med. J. Austr.*, 1, 3, 1980, pp. 136–7.

Diabetes and other metabolic disorders

S. S. Ajgaonkar, 'Yoga therapy – its place in treatment of diabetics and obesity', *J. Diab. Assoc. Ind.*, 18, 2, 1978, pp. 75–8.

M. V. Divekar and M. T. Mulla, 'Effect of yoga therapy in diabetes and obesity', in *Clinical Diabetes Update*, Diabetic Association of India, 1981, pp. 40–2.

M. L. Gharote, 'An evaluation of the effects of yogic treatment on obesity – a report', *Yoga Mimamsa*, 19, 1, 1977, pp. 13–37.

P. V. Karambelkar, 'Care by a diabetic through yogic techniques', *Yoga Mimamsa*, 18, 3 and 4, 1976, pp. 79–88.

U. Mahamood, *Yoga and Diabetes*, report submitted to the Diabetic Association of India by the Department of Medicine and Diabetology, Stanley Hospital, Madras, 1979–80.

A. Maini *et al.*, 'Effect of yoga on blood sugar and serum cholesterol levels', *Ind. Med. Gaz.*, 114, 1980, p. 114.

B. Venkat Rao, 'Report on the diabetic research of the government of India', *Prakriti*, December, 1976, p. 33.

P. S. Rugmini and R. N. Sinha, 'The effect of yoga therapy in diabetes mellitus', in *Proceedings of a Seminar on Yoga, Science and Man*, Central Council of Research in Indian Medicine and Homeopathy, New Delhi, 1975, pp. 175–81.

A. G. Shembekar *et al.*, 'Yogic exercises in management of diabetes mellitus', *J. Diab. Assoc. Ind.*, 20, 3, 1980, p. 167.

T. H. Tulpule, 'Yogic exercises and diabetes mellitus. Madumeha', *J. Diab. Assoc. Ind.*, 17, 1977, pp. 37–8.

Drugs

H. Benson, 'Decreased alcohol intake associated with the practice of meditation. A retrospective investigation', *Ann. N.Y. Acad. Sci.*, 233, 1974, pp. 174–7.

H. Benson *et al.*, 'Yoga for drug abuse', *New England J. Med.*, 281, 1969, p. 1133.

E. Brautigam, *The Effect of Transcendental Meditation on Drug Abuses*, City of Malmo, 1971.

M. Shafie, R. Lavely and R. Jaffe, 'Meditation and the prevention of alcohol abuse', *Am. J. Psychiatry*, 132, 1975, pp. 942–5.

Gastrointestinal disorders

B. P. Desai and M. V. Bhole, 'Gastric responses to vastra dhauti and standard test meal – a comparison', *Yoga Review*, 2, 1, 1982, pp. 53–8.

M. L. Gharote, 'Effect of air swallowing on the gastric acidity – a pilot study', *Yoga Mimamsa*, 14, 1 and 2, 1971, pp. 7–10.

M. L. Gharote and P. V. Karambelkar, 'Influence of danda dhauti on gastric acidity – a preliminary communication', in S. Digambarji (ed.), *Collected Papers on Yoga*, Kaivalyadhama, Lonavla, 1975, pp. 41–7.

S. P. Mishra and R. H. Singh, 'Effect of certain yogic asanas on the pelvic congestion and its anatomy', *Ancient Science Of Life*, 4, 2, 1984, pp. 127–8.

M. N. Parandekar, 'Constipation – its causes and cure', *Yoga Mimamsa*, 4, 4, 1933, pp. 332–7.

R. A. Phillips *et al.*, 'A new approach to the study of gastrointestinal functions in man by an oral lavage method', *Chinese M. J.*, 23, 1976, pp. 85–95.

Jan Erick Sigdell, 'An ancient method of kriya yoga – a modern alternative to the artificial kidney', *Indian Review*, October, 1979, pp. 38–44.

R. H. Singh and K. N. Udupa, 'Role of certain yogic practices in the prevention and treatment of gastrointestinal disorders', *J. Res. Med. Yoga and Homeo.*, 11, 2, June, 1976, pp. 51–9.

R. G. Smith, 'Whole gut irrigation – a new treatment for constipation', *Br. Med. J.*, 2, 1978, pp. 396–7.

E. Wittkower and K. Dhawan, 'Treatment of chronic functional constipation with the methods of yoga practice' (German), *Deutsche Medizinische Wochenschrift*, 59, 1933, pp. 284–5.

T. K. Young and S. C. Lee, 'Gastrointestinal dialysis in the therapy of uremia', *Kidney International*, 13, Suppl. 8, S-185-A, 1978, p. 187.

T. K. Young, S. C. Lee and C. K. Tang, 'Diarrhoea therapy of uremia', *Clinical Nephrology*, 11, 2, 1979, pp. 86–91.

Hypertension and other cardiovascular disorders

R. C. Agrawal *et al.*, 'Effect of savasana on vascular response to cold pressure test, serum cholesterol level and platelet stickiness', *Ind. Heart J.*, 29, 4, 1977, pp. 182–5.

L. R. Bali, 'Long-term effect of relaxation on blood pressure and anxiety level of essential hypertensive males – a controlled study', *Psychosom. Med.*, 46, 8, 1979, pp. 637–45.

C. Barr Taylor *et al.*, 'Relaxation therapy and high blood pressure', *Arch. Gen. Psychiat.*, 34, 1977, p. 339.

N. A. Belaia, 'Effect of certain asanas used in the system of yoga on the central nervous and cardiovascular system' (Russian), *Vopr. Kurortol Fizioter Fiz. Kult.*, 0, 3, May–June, 1976, pp. 13–18.

H. Benson, 'Systemic hypertension and the relaxation response', *New England J. Med.*, 296, 20, 1977, pp. 1152–6.

H. Benson *et al.*, 'Decreased systolic blood pressure in hypertensive subjects who practised meditation', *J. Clin. Invest.*, 52, 1973, p. 8.

H. Benson, B. A. Rosner and B. R. Marzetta, 'Decreased blood pressure in borderline hypertensive subjects who practised meditation', *J. Chronic Dis.*, 27, 1974, pp. 163–9.

W. Brace, 'The effects of yoga on blood pressure and anxiety', *Yoga Journal*, 1, 1976, p. 6.

K. K. Datey, 'Shavasana and biofeedback in the management of hypertension', paper read at International Conference of Yoga and Psychic Research, Bangalore, 1977.

K. K. Datey, S. Deshmukh, C. Dalvi and S. L. Vinekar, 'Shavasana: a yogic exercise in the management of hypertension', *Angiology*, 20, 1969, pp. 325–33.

H. L. Deabller, E. Fedel and R. L. Dillenkoffer, 'The use of relaxation and hypnosis in lowering high blood pressure', *Am. J. Clin. Hypn.*, 16, 1973, p. 75.

B. P. Desai and M. L. Gharote, 'Effect of kapalabhati on some blood constitutents', in *Abstracts of the 53rd Annual Meeting of the Society of Biological Chemists (India)*, Dept. of Biochemistry, V.P. Chest Institute, University of Delhi, October, 1984, p. 20.

C. Dostalek and V. Lepicovska, 'Hathayoga – a method for prevention of cardiovascular diseases,' *Activ. Nerv. Sup.*, 24, Suppl. 3, 1982, pp. 444–52.

A. Friedell, 'Automotive attentive breathing in angina pectoris', *Minnesota Medicine*, 31, 1948, pp. 875–81.

H. Gaertner, L. Gaertner, W. Goszcz and T. Pasek, 'Influence of sirsasana-head stand posture of 30 min. duration on blood composition and circulation' (Polish), *Acta Physiol. Pol.*, 16, 1, 1965, p. 44.

M. Gayer, 'The position of concentrated relaxation in a training programmer for the psycho prevention of myocardial infarction', *Psychia. Neuro. Med. Psychol. Leips.*, 27, 9, 1975, pp. 542–9.

C. Lakshmikanthan *et al.*, 'Long term effects of yoga on hypertension and/or coronary artery disease', *J. Ass. Phys. India*, 27, 1979, p. 1055.

V. Lepicovska, C. Dostalek and M. Vlcek, 'Vascomotor changes effected by breathing manoeuvres', *Activ. Nerv. Sup.*, 25, 3, 1983, pp. 195–6.

K. Nespor, 'Yoga and cardiovascular disease prevention' (Czech), *Cas. Lek. Cesk.*, 118, 4, 1979, pp. 333–5.

R. C. Pandey, V. M. Bhatnagar and R. K. Kalra, 'Attenuation of cardiac vulnerability to dysrhythmias by exploitation of cardiovascular reflexogenicity through uddiyan and jalandhar bandhas', paper read at All India Conference on Yoga and Its Integration in Modern Education, Indian Institute of Technology, Kanpur, 10–13 September, 1981.

C. H. Patel, 'Yoga and biofeedback in the management of hypertension', *Lancet*, 2, 1973, p. 1053.

C. Patel, M. Marmot, D. J. Terry, M. Carruthers, and P. Sever, 'Coronary risk factor reduction through biofeedback, aided relaxation and meditation', *Circulation*, 60, 1979, p. 226.

R. K. Peters, H. Benson and J. M. Peters, 'Daily relaxation response breaks in a working population: 2. Effects on blood pressure', *Am. J. Public Health*, 67, 10, 1977, pp. 954–9.

D. P. Redmond, M. S. Gaylor and R. H. McDonald, 'Blood pressure and heart rate response to verbal instructions and relaxation in hypertension', *Psychosom. Med.*, 36, 4, 1974, pp. 285–97.

D. Reinharez, 'Value of yoga in phlebology' (French), *Phlebology*, 21, 1968, pp. 147–51.

P. Roberti and V. Stanghellini, 'Recenti acquisizione in medicina psicosomatics: ipertensine arteriosa e yogaterapia', *Riv. Neuropsichiat. Sci. Aff.*, 1, 1980, p. 48.

P. Seep *et al.*, 'Meditation training and essential hypertension: a methodological study', *J. Behav. Med.*, 3, 1, 1980, pp. 59–71.

J. E. Shoemaker *et al.*, 'The effects of muscle relaxation on blood pressure of essential hypertensives', *Behav. Res. and Therapy*, 13, 1, 1976, pp. 29–43.

S. B. Shukla, I. D. Saxena, A. Kumar and Rajani Shukla, 'Is pranayama a cardiac exercise as well?', *J. Res. Ind. Med. Yoga and Homeo.*, 14, 3–4, 1979, pp. 146–8.

R. A. Stone and J. de Leo, 'Psychotherapeutic control of hypertension', *New England J. Med.*, 295, 1976, pp. 80–4.

C. B. Taylor, J. W. Farquar, E. Nerson and S. Agros, 'Relaxation therapy and high blood pressure', *Arch. Gen. Psychiat.*, 34, 3, 1977, p. 339.

T. H. Tulpule, Shantilal J. Shah and H. K. Havelivala, 'Yogic exercises in the management of ischaemic heart disease', *Indian Heart Journal*, 23, 4, 1973, pp. 259–64.

T. H. Tulpule *et al.*, 'A method of relaxation for rehabilitation after myocardial infarction', *Indian Heart Journal*, 32, 1, 1980, pp. 1–7.

Pedrodo Vicente, M. L. Gharote and J. M. Bhagwat, 'Effect of kapalabhati and uddiyana bandha on cardiac rhythm', *Yoga Mimamsa*, 23, 1, 1984, pp. 44–62.

Neurological conditions

D. K. Gosavi, 'Scope of yoga in the treatment of vertigo', paper read at VIII International Conference of Neuro-otological and Equilibrometric Society of India, Bombay, November, 1981.

Anthony D. Heyes, 'Blindness and yoga', *A New Outlook for the Blind*, 68, 9, 1974, pp. 385–93.

M. G. Mokashi, 'Yogasanas as key postures in cerebral palsy', *J. Ind. Assoc. Physiotherapists*, 1, 3, 1973, pp. 27–31.

R. H. Singh, R. M. Shettiwar and K. N. Udupa, 'Role of some hathayogic practices in the management of chronic rheumatic diseases', in *Proceedings of the IX National Seminar on Rheumatic Disease*, MML Centre for Rheumatic Disease, New Delhi, 1978.

L. Stejskal, 'Respiration rhythm and variation of neuromuscular excitability' (Czech), *Cas. Lek. Ces.*, 107, 1968, pp. 1551–5.

Pain

H. Benson *et al.*, 'The usefulness of relaxation response in the therapy of headache', *Headache*, 14, 1974, pp. 49–52.

S. Brena, 'Chronic pain: a point of contact between yoga teaching and present Western thought', paper read at IV Cong. Int. Coll. Psychosom. Med., Kyoto, 1977.

H. C. Burry, 'Athletic fitness: its role in prevention of accidents and injuries', *Practitioner*, 206, 1971, pp. 227–33.

Herbert A. de Vries, 'Prevention of muscular distress after exercise', *Research Quarterly*, 32, 1961, pp. 177–85.

Herbert A. de Vries, 'Electromyographic observations of the effects of static stretching upon muscular distress', *Research Quarterly*, 32, 1961, pp. 468–79.

T. Tanaka, J. R. M. Goyeche, T. Noda, N. Mishima, Y. Ohno and Y. Ikemi, 'A case study of migraine alleviation by means of yoga', paper read at IV Cong. Int. Coll. Psychosom. Med., Kyoto, 1977.

G. Warner, 'Relaxation therapy in migraine and chronic tension headache', *Med. J. Austr.*, 1, 10, 1975, pp. 298–301.

S. Weller, 'Yoga for tired legs and aching back', *Can. Nurse*, 73, 5, 1977, pp. 20–3.

Stress

Sharmila Agarwal, S. R. Pathak, A. K. Sharma and K. N. Udupa, 'A preliminary study of neurophysiological changes following meditation practices in stress disorders', in *Abstracts of the 53rd Annual Meeting of the Society of Biological Chemists (India)*, Dept. of Biochemistry, V.P. Chest Institute, University of Delhi, October 1984, p. 84.

D. J. Goleman and G. E. Schwartz, 'Meditation as an intervention in stress activity', *J. Clin. Consult. Psychology*, 44, 1976, pp. 456–66.

M. Horvath *et al.*, 'Seminar on primary prevention of psychosomatic disorders in relation to work stress combined with training in relaxation techniques', *Vnitr. Lek.*, 25, 1, 1979, pp. 89–92.

P. V. Karambelkar, M. V. Bhole and M. L. Gharote, 'Effect of yogic asanas on uropepsin excretion', *Ind. J. Med. Res.*, 57, 5, 1969, pp. 944–7.

P. V. Karambelkar, M. L. Gharote, S. K. Ganguly and A. M. Moorthy, 'Cholesterol level and yogic training programme', *J. Res. Ind. Med. Yoga and Homeo.*, 13, 4, 1978, pp. 1–6.

A. S. Sethi, 'Using meditation in stress situation', *Dimens. Health Serv.*, 57, 1, 1980, pp. 24–8.

YOGA AND PSYCHIATRY

T. Candelent *et al.*, 'Teaching transcendental meditation in a psychiatric setting', *Hospital and Community Psychiatry*, 26, 3, 1975, pp. 156–9.

D. K. Deshmukh, 'Experiences in management of psychiatric and psychosomatic disorders with yoga', *J. Yoga Inst.*, 18, December, 1972, p. 5.

E. Gellhorn, 'Central nervous system tuning and its implications for neuropsychiatry', *J. Nerv. Ment. Dis.*, 147, 1968, p. 148.

B. C. Glueck and C. S. Stroebel, 'The use of transcendental meditation in a psychiatric hospital', paper read at 127th Annual Meeting of the American Psychiatric Association, Detroit, May, 1974.

B. C. Glueck and C. F. Stroebel, 'Biofeedback and meditation in the treatment of psychiatric illness', *Comprehensive Psychiatry*, 15, 1975, pp. 305–21.

J. Malhotra, 'Yoga and psychiatry: a review', *J. Neuropsychiatry*, 4, 1963, pp. 375–85.

N. Vahia, S. Vinekar and D. Doongaji, 'Some ancient Indian concepts in the treatment of psychiatric disorders', *Br. J. Psychiatry*, 112, 1966, pp. 1089–96.

Alan W. Watts, 'Asian psychology and modern psychiatry', *Am. J. Psychoanalysis*, 13, 1, 1953, pp. 25–30.

I. Wigley, 'Community psychiatric nursing: yoga as therapy', *Nursing Times*, 72, 44, 1976, pp. 1716–17.

YOGA AND PSYCHOANALYSIS

H. S. Brar, 'Yoga and psychoanalysis', *Br. J. Psychiatry*, 116, 1970, pp. 201–6.

P. Carrington and H. S. Ephron, 'Meditation and psychoanalysis', *J. Am. Acad. Psychoanal.*, 3, 1, 1975, pp. 43–57.

A. Chakraborty, 'Yoga and psychoanalysis', *Br. J. Psychiatry*, 117, 1970, p. 478.

M. Choisy, *Yoga et psychoanalyse*, Ed. du Mont Blanc, Paris, 1945.

S. B. Harchand, 'Yoga and psychoanalysis', *Br. J. Psychiatry*, 116, 1970, p. 201.

J. S. Negi, 'Yoga and psychoanalysis', *Comprehensive Psychiatry*, 8, 1967, p. 160.

E. Servadio, 'Psychoanalysis and yoga', *Med. Bull.*, 8, 1940, p. 1.

YOGA AND PSYCHOTHERAPY

Swami Ajaya, *Psychotherapy East and West, A Unifying Paradigm*, Himalayan Institute of Yoga Science and Philosophy, Chicago, 1983.

P. C. Boswell *et al.*, 'Effects of meditation on psychological and physiological measures of anxiety', *J. Consult. Clin. Psychol.*, 47, 1979, p. 606.

H. H. Dawley, 'Anxiety reduction through self-administered relaxation', *Psychological Reports*, 36, 2, 1975, pp. 595–7.

D. K. Deshmukh, 'Yoga in management of psychoneurotic, psychotic and psychosomatic conditions', *J. Yoga Inst.*, 16, 10, 1971, pp. 154–8.

K. R. Dhawan, 'Yoga und seine Psychologischen Bedautungen', *Medizinische Klinik*, 1956, pp. 2231–3.

Philip C. Ferguson and John C. Gowan, 'The influence of transcendental meditation on anxiety, depression, aggression, neuroticism and self-actualisation', paper given at the California State Psychological Association, Fresno, California, 1974

M. Girodo, 'Yoga meditation and flooding in the treatment of anxiety neurosis', *J. Behav. Ther. and Exp. Psychiat.*, 5, 1974, p. 157.

J. R. M. Goyeche, 'Yoga and psychotherapy', *Yoga Journal*, 1, 1976, p. 6.

L. Gresham and J. M. Stedman, 'Oriental defence exercises as reciprocal inhibitors of anxiety', *J. Behav. Ther. and Exp. Psychiat.*, 2, 1971, pp. 117–19.

Paul Grim, 'Psychotherapy by somatic alteration', *Mental Hygiene*, 53, 1969, pp. 451–8.

S. Grinblat, 'Shavasana, autopsychorelaxation', *Semana Medica*, 115, 1959, pp. 928–30.

Gustav R. Heyer, 'Yoga und psychotherapie', *Jahrbuch fur Psychologie und Psychotherapie*, 6, 1958, pp. 350–5.

D. F. Hutchings *et al.*, 'Anxiety management and applied relaxation in reducing general anxiety', *Behav. Res. Ther.*, 18, 1980, p. 181.

Y. Ikemi *et al.*, 'The biologic wisdom of self-regulatory mechanism of normalisation in autogenic and oriental approaches in psychotherapy', *Psychother. Psychosom.*, 25, 1–6, 1975, pp. 99–108.

H. C. Kocher and V. Pratap, 'Neurotic trend and yogic practices', *Yoga Mimamsa*, 14, 1 and 2, 1971, pp. 34–40.

H. C. Kocher and V. Pratap, 'Anxiety level and yogic practices', *Yoga Mimamsa*, 15, 1, 1972, pp. 11–16.

A. Kondo, 'Zen in psychotherapy: the virtue of sitting', *Chicago Rev.*, 12, 1958, pp. 57–64.

W. Kretschmer, 'Meditative techniques in psychotherapy', *Psychologia*, 5, 1962, pp. 76–83.

M. Lerner, 'Yoga concentration and psychotherapy', *Acta Psychiat. Am. Lat.*, 17, 1971, p. 410.

A. Meares, 'The relief of anxiety through relaxing meditation', *Austr. Fam. Phys.*, 5, 1976, p. 906.

R. Meyer, 'The practice of awareness as a form of psychotherapy', *J. Religious Health*, 10, 1971, pp. 333–45.

T. Pasek, C. Nowakowska, B. Fellman, J. Hauser and A. Sluzewska, 'The evaluation of yoga-type relaxation concentration training on patients with psychogenic mental disturbances', in *The 6th World Congress of the International College of Psychosomatic Medicine*, International College of Psychosomatic Medicine, Montreal, 1981, p. 129.

T. Pasek and W. Romanowski, 'Role of steered rhythms in the prevention of psychoneurological disturbances' (Polish), *Med. Lot.*, 38, 1972, pp. 133–5.

P. Pitta *et al.*, 'Cognitive stimulus control programme for obesity with emphasis on anxiety and depression reduction', *Int. J. Obes.*, 4, 1980, p. 227.

H. B. Puryear *et al.*, 'Anxiety reduction associated with meditation: home study', *Percept. Mot. Skills*, 42, 1976, p. 527.

G. Savermann, 'Psychotherapy and yoga', *Zeit. fur Psichosom. Med. und Psychoanalyse*, 26, 4, 1980, p. 364.

D. Thomas *et al.*, 'Comparison of transcendental meditation and progressive relaxation in reducing anxiety', *Br. Med. J.*, 2, 1978, p. 172.

N. S. Vahia, 'Psychophysiologic therapy based on the concepts of Patanjali: a new approach to the treatment of neurotic and psychosomatic disorders', *Am. J. Psychother.*, 27, 1973, pp. 557–65.

Appendix 3 *Useful Addresses*

AUSTRALIA

**International Yoga Teachers'
Association**
PO Box 207
St Ives
NSW 2075

**Physiotherapy Centre for
Maternal and Child Health**
347–349 Riley Street
Surry Hills
NSW 2010

BRITAIN

British Wheel of Yoga
Grafton Range
Grafton
N. Yorks YO5 9QQ

Natraj Yoga Centre
46 Crouch Hall Road
London N8 8HJ

Servite House
Queen's Walk
Ealing
London W5 1TL

Yoga for Health Foundation
Ickwell Bury
Northill
Nr Biggleswade
Bedfordshire SG18 9EF

**National Federation of Spiritual
Healers**
Old Manor Farm Studio
Church Street
Sunbury-on-Thames
Middlesex TW16 6RG

CZECHOSLOVAKIA

**Ceskoslovenska Academie Ved
Ustav Fysiolockych Regulaci**
Bulovka, Pavilion 11
18085 Praha-8

INDIA

**GS College of Yoga
Kaivalyadhama**
Lonavla 410 403

Kaivalyadhama SMYM Samiti
Lonavla 410 403

**Kaivalyadhama ICY Health
Centre**
43 Netaji Subhash Road
Marine Drive
Bombay 400 002

ITALY

International Association of Ayurveda, Naturopathy and Yoga
Vila Era
Via Rivetti 61
130 69 Vigliano Biellesse (Vc)

JAPAN

Institute for Religious Psychology
4-11-17 Inokashira
Mitaka-shi
Tokyo 181

SPAIN

Institute of Relaxation Technique and Yogic Therapy
Dept of Physiology
University of Seville

USA

The Evolving Institute
1408 Pine Street
Boulder
Colorado 80302

Index